A Three-Component Hybrid Image Compression Algorithm

Michael McNees

A Three-Component Hybrid Image Compression Algorithm

The Development & Analysis of a 3-Way Hybrid Image Compression Method Incorporating Fractal, DCT JPEG), and Wavelet (JPEG2000) Components

VDM Verlag Dr. Müller

Impressum/Imprint (nur für Deutschland/ only for Germany)

Bibliografische Information der Deutschen Nationalbibliothek: Die Deutsche Nationalbibliothek verzeichnet diese Publikation in der Deutschen Nationalbibliografie; detaillierte bibliografische Daten sind im Internet über http://dnb.d-nb.de abrufbar.

Alle in diesem Buch genannten Marken und Produktnamen unterliegen warenzeichen-, marken- oder patentrechtlichem Schutz bzw. sind Warenzeichen oder eingetragene Warenzeichen der jeweiligen Inhaber. Die Wiedergabe von Marken, Produktnamen, Gebrauchsnamen, Handelsnamen, Warenbezeichnungen u.s.w. in diesem Werk berechtigt auch ohne besondere Kennzeichnung nicht zu der Annahme, dass solche Namen im Sinne der Warenzeichen- und Markenschutzgesetzgebung als frei zu betrachten wären und daher von jedermann benutzt werden dürften.

Coverbild: www.purestockx.com

Verlag: VDM Verlag Dr. Müller Aktiengesellschaft & Co. KG
Dudweiler Landstr. 99, 66123 Saarbrücken, Deutschland
Telefon +49 681 9100-698, Telefax +49 681 9100-988, Email: info@vdm-verlag.de

Herstellung in Deutschland:
Schaltungsdienst Lange o.H.G., Berlin
Books on Demand GmbH, Norderstedt
Reha GmbH, Saarbrücken
Amazon Distribution GmbH, Leipzig
ISBN: 978-3-639-17612-4

Imprint (only for USA, GB)

Bibliographic information published by the Deutsche Nationalbibliothek: The Deutsche Nationalbibliothek lists this publication in the Deutsche Nationalbibliografie; detailed bibliographic data are available in the Internet at http://dnb.d-nb.de .

Any brand names and product names mentioned in this book are subject to trademark, brand or patent protection and are trademarks or registered trademarks of their respective holders. The use of brand names, product names, common names, trade names, product descriptions etc. even without a particular marking in this works is in no way to be construed to mean that such names may be regarded as unrestricted in respect of trademark and brand protection legislation and could thus be used by anyone.

Cover image: www.purestockx.com

Publisher:
VDM Verlag Dr. Müller Aktiengesellschaft & Co. KG
Dudweiler Landstr. 99, 66123 Saarbrücken, Germany
Phone +49 681 9100-698, Fax +49 681 9100-988, Email: info@vdm-publishing.com

Printed in the U.S.A.
Printed in the U.K. by (see last page)
ISBN: 978-3-639-17612-4

To my wife and son,
DeVonna and Joseph Darrell

LIST OF ABBREVIATIONS

AC	Alternating Current
CR	Compression Ratio
dB	Decibel
DC	Direct Current
DCT	Discrete Cosine Transform
DPCM	Differential Pulse Code Modulation
DWT	Discrete Wavelet Transform
FDCT	Forward Discrete Cosine Transform
IDCT	Inverse Discrete Cosine Transform
IFS	Iterated Function System
JPEG	Joint Photographic Experts Group
MSE	Mean Square Error
PIFS	Partitioned Iterated Function System
PSNR	Peak Signal to Noise Ratio
Qs	Explicit Quantization Step Size for JPEG2000
RMS	Root Mean Square
RMSE	Root Mean Square Error

ACKNOWLEDGEMENTS

I would like to express my sincere gratitude and thanks to Dr. William Stapleton for all of his guidance, support and supervision of this research project. I also want to express my love and appreciation to my parents. Without their continuing love and support, returning to school would have only been a dream, and none of this work would have been possible. In closing, I would also like to thank the rest of my friends and family, whose names I have not mentioned here, for their support, as each has played an important role in my continuing success.

CONTENTS

LIST OF TABLES

LIST OF FIGURES

CHAPTER 1
INTRODUCTION

Every day information is processed, stored, and transmitted digitally. With rapid advancements in multimedia technologies, the use of digital images by individuals is quickly increasing. This increase includes not only the number of digital images used but also the resolution of these images. Image compression addresses the issue of reducing the amount of data required to represent a digital image. The basis of this reduction is in the removal of redundant data from the image [1].

Image compression techniques are classified as being either lossless or lossy. The current lossy data compression methods are acceptable for non-critical applications and more often used because of the greater compression yield versus the amount of error in the resulting image. These methods typically control the amount of error introduced in the resulting image by adjusting the compression level. However, as the level of image compression increases, so too does the amount of error. Lossy image compression techniques offer the advantage of yielding high compression ratios (CR). One disadvantage, though, is that they only store an approximation to the original image, making them undesirable for use in critical applications.

Since the mid-1980s, members from the International Organization for Standardization (ISO) and the International Telecommunication Union (ITU) have been collaborating to devise a joint international standard for the compression of multilevel, grayscale and color, still frame images. This collaborative effort is known as JPEG, or the Joint Photographic Experts Group [2]. The algorithm devised by and commonly known as JPEG is currently the industry-wide standard, but alternative compression algorithms are being explored. One of these methods is fractal image compression.

Fractal image compression is a radical departure from conventional compression techniques in that it partitions an image into blocks and uses contractive mapping to map range blocks to domains to form a Partitioned Iterated Function System (PIFS) [3]. In other words, rather than storing data for each individual pixel, formulas or instructions are stored to create the image. Traditional methods try to identify similarities among the image elements by searching the entire image pixel space. Thus, the encoding involved in this traditional method often leads to very high computational complexity and extremely long execution times.

JPEG image compression takes advantage of the fact that a certain amount of data loss is acceptable. Local approximation is the main mathematical and physical theme behind the JPEG algorithm. In comparison with fractal compression, JPEG is very efficient at encoding low-contrast blocks whereas fractal methods exhibit greater efficiency when dealing with high-contrast blocks. For this reason, these two methods tend to be complimentary to each other when dealing with which image elements each method compresses best.

A newer method developed to provide superior image quality performance is the JPEG2000 algorithm. This method also provides functionality that current standards either do not address efficiently or do not address at all. By taking advantage of new technologies, the JPEG2000 standard provides features of vital importance for many high-end applications [2]. Without sacrificing performance, this coding method should provide low bit-rate operation and image quality performance superior to existing standards. In this book, a method for implementing two hybrid image compression algorithms will be explored. The first will encompass both a searchless fractal image compression algorithm and the JPEG image compression algorithm. The second will encompass a searchless fractal image compression algorithm, a JPEG image compression algorithm, and also a JPEG2000 image compression algorithm. The resulting hybrid algorithms should exceed the performance of any of the three methods alone when comparing the relationships of the resulting image compression ratio with introduced error.

1

CHAPTER 2
FRACTALS

2.1 Fractal Image Compression Algorithm

Fractal image compression is quite different from other image compression methods. Unlike JPEG compression, which quantizes the entire frequency spectrum for the image of interest, fractal compression makes use of the self-similarity that exists within an image at different scales. As an example, consider a blue sky in an image, which consists of smaller sections of blue, or consider the branches on a tree. These branches can be subdivided into smaller and smaller branches, all having similar structure.

Both JPEG and fractal compression techniques are considered lossy and eliminate duplicate information within an image in order to represent it with a reduced data set. As opposed to storing individual pixel data, fractal compression stores instructions or mathematical formulas for creating an image, making the images compressed by this method resolution independent. This allows scaling of the images to resolutions higher than that of the original image, thus making them well suited for graphics systems composed of devices with differing resolutions such as printers and graphics cards. However, a tradeoff exists between compression time and image quality with this technique.

The scheme behind fractal image compression works by partitioning an image into blocks and using contractive mapping to map range blocks to domains [3]. Fractal compression involves encoding the original image by finding some finite set of mathematical equations that can accurately describe the image. Fractal image coding techniques also assume that all images contain some self-similarity. Typically, the more self-similarity, the greater the compression ratio is going to be for that image.

2.2 Iterated Function Systems (IFS)

The current methods employed in fractal image compression are based on the theory describing Iterated Function Systems (IFS), much of which is attributed to Hutchinson [5]. This theory describes the types of functions that perform transformations on a data set to map points within the data set, in particular those which, when iterated, tend to converge to a single fixed point. When applied to images, transform functions can be thought of as those which map pixels from one part of an image plane to another, so that, when iterated, produce a fixed point data set. Although IFS's can yield extremely high compression factors, the cost of compressing the image can be large as well. One advantage of IFS though is that images possessing substantial geometric regularity are the ones that are compressed, whereas those that look like noise are more likely to be expanded [13]. The Sierpinsky Triangle, shown in Figure 2.1, is a well-known example of a simple set of iterated functions. Contractivity is one of the requirements for the transformations and without it the iterative process would be unstable.

Invented by Michael F. Barnsley [6], an IFS can be represented by a collection of functions that use the general form of Equation (2.1). Values X_n and Y_n are determined by performing the transformation w on the current values for x and y. The w transformation coefficients are represented by a, b, c, and d, which determine the scale, position, and orientation for the resulting data. Values e and f aid in performing a linear translation of the array [7]. When combined with other techniques, IFS are excellent image compression tools. The resulting images are stored as a collection of transformations instead of a collection of pixels. In constructing an image in this manner, objects naturally occurring in nature, such as fern leaves, clouds, and mountains are easily produced. Transformations involving rotation, translation, scaling, and skewing of an original image are called affine transformations. An affine transformation is the most general linear transformation that can be performed on an image. Affine transformations remap any coordinate system in a plane into another coordinate system by a combination of translation, shear, and scaling. Each iteration through an IFS collection as shown in Equation (2.1) will transform the image, producing more detail on an increasingly smaller scale. IFS theory uses repeated iterations of the affine transformations of the plane to reproduce a fractal image.

$$\begin{pmatrix} X_n \\ Y_n \end{pmatrix} = w \begin{pmatrix} x \\ y \end{pmatrix} = \begin{pmatrix} a & b \\ c & d \end{pmatrix} \begin{pmatrix} x \\ y \end{pmatrix} + \begin{pmatrix} e \\ f \end{pmatrix}$$

(2.1)

$$\begin{pmatrix} X_n \\ Y_n \\ Z_n \end{pmatrix} = w \begin{pmatrix} x \\ y \\ z \end{pmatrix} = \begin{pmatrix} a & b & 0 \\ c & d & 0 \\ 0 & 0 & s \end{pmatrix} \begin{pmatrix} x \\ y \\ z \end{pmatrix} + \begin{pmatrix} e \\ f \\ O \end{pmatrix}$$

2.3 Sierpinski Triangle

The Sierpinski Triangle, shown in Figure 2.1, is an example of an IFS collection on an arbitrary image. By scaling, copying, and remapping some arbitrary initial image, an entirely new image known as the Sierpinski Triangle will be produced. This image, called an attractor, is the outcome from multiple iterations of an IFS collection. The resulting image will converge to the attractor for some unique set of IFS values.

Figure 2.1 Sierpinski Triangle with $n = 8$ iterations.

For those images that converge, regardless of how much someone zooms in on that image, detail will still be viewable. Although the concept of whole image self-similarity does not apply to all images, it is true that portions of an image are similar to other portions of the same image [4]. Taking only a small portion of the image in Figure 2.1, it becomes possible, by using affine transformations, to map one part of the image into another. Exploiting the self-similarity that exists naturally in most images makes it possible to represent an image by a collection of IFS sets instead of a pixel array. The three affine transformations shown in Equation (2.2) are needed to represent the algorithm for Sierpinski's Triangle as mathematical functions.

$$w_1 \begin{bmatrix} x \\ y \end{bmatrix} = \begin{bmatrix} \frac{1}{2} & 0 \\ 0 & \frac{1}{2} \end{bmatrix} \begin{bmatrix} x \\ y \end{bmatrix} + \begin{bmatrix} 0 \\ 0 \end{bmatrix}$$

$$w_2 \begin{bmatrix} x \\ y \end{bmatrix} = \begin{bmatrix} \frac{1}{2} & 0 \\ 0 & \frac{1}{2} \end{bmatrix} \begin{bmatrix} x \\ y \end{bmatrix} + \begin{bmatrix} \frac{1}{2} \\ 0 \end{bmatrix}$$

(2.2)

$$w_3 \begin{bmatrix} x \\ y \end{bmatrix} = \begin{bmatrix} \frac{1}{2} & 0 \\ 0 & \frac{1}{2} \end{bmatrix} \begin{bmatrix} x \\ y \end{bmatrix} + \begin{bmatrix} \frac{1}{4} \\ \frac{1}{2} \end{bmatrix}$$

3

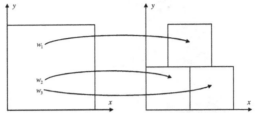

Figure 2.2 Sierpinski Triangle mapping from the three transformations.

This process can be viewed as three transformations mapping from the original triangle image to the new triangle image as shown in Figure 2.2. It is important to notice that there is little difference between the fifth and sixth iterations as shown in Figure 2.3. The MATLAB code demonstrating the implementation of Sierpinski's Triangle is provided in Appendix A. The Sierpinski's Triangle sequence visually converges to one triangle within a certain threshold. Conversely, if we begin the sequence backwards, for the Sierpinski triangle, the only thing we need to know are the mappings w_1, w_2 and w_3. In essence, this is how fractal image compression works. Given any image, this IFS algorithm will generate Sierpinski's Triangle, as illustrated in Figure 2.4.

n=0 iterations   n=1 iterations   n=5 iterations   n=6 iterations

Figure 2.3 Sierpinski Triangle iteration generation sequence.

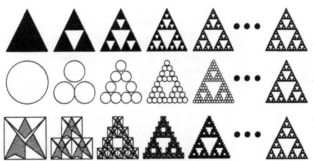

Figure 2.4 Iterations of Sierpinski's Triangle generated from different initial images [24].

2.4 Partitioned Iterated Function System (PIFS)

Jacquin [8, 9] is credited with the first automated attempt at creating an IFS from an ordinary image. His idea was to first partition the image into non-overlapping areas called ranges and areas that might overlap called domains. The union of these ranges comprises the original image while the union of the domains might not comprise the original image. In other words, both sets cover the entire image space, with the domain set allowing overlapping. As shown in Figure 2.5, ranges are represented by small blocks and domains by large blocks. The process of encoding an image consists of determining the

4

best possible affine transformations by searching partial or global domain block spaces. As a shortcoming of using affine transformations, the partition schemes for the domain set and the range set have to give the same geometric shaped domain and range blocks, which are usually squares or rectangles. From Jacquin's original scheme, which is widely accepted in fractal image compression, the domain blocks are set to be twice as big as the range blocks. The reason for allowing the domains to overlap is to smooth artifacts between the blocks in the decoding process. The mapping between the domain and the range blocks is shown in Figure 2.6. For each range block, we need to find a proper domain block to map to. The final map set is composed of mappings for each range block from the range set. This is the basis behind the idea of Partitioned Iterated Function Systems (PIFS) that Jacquin introduced as part of his dissertation in 1989.

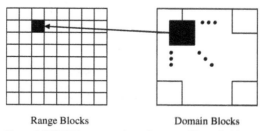

Range Blocks   Domain Blocks

Figure 2.5 PIFS Representation---Range and Domain blocks.

PIFS is recognized as a significant improvement over IFS. It reduces the amount of search time both theoretically and technically. Furthermore, it provides some possible aspects of improving fractal encoding like using different partitioning schemes of taking different mapping methods. A PIFS is a generalization of an IFS, and attempts to simplify the IFS computations by partitioning the whole space into subspaces. In other words, PIFS is a restricted version of IFS. A shortcoming of this approach is that the matching process is time consuming and computationally intensive for large domain spaces.

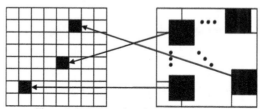

Figure 2.6 Mapping domain sets to range sets.

2.5 Conclusions

Fractal compression methods have both advantages and weaknesses when compared to other image compression techniques. Perhaps the biggest advantage is the ability to achieve relatively high compression ratios within a certain acceptable window of recovery. This is largely due to the fractal encoding process. In this process, an attempt is made to find close mappings, or affine transformations, for each range block from the domain blocks. The only information that needs to be stored is the domain location along with the scaling and offset coefficients for each transformation.

The success of fractal image compression depends largely on the self-similarity exhibited in the image space. A big drawback to this is there is no guarantee that the probability of matching range and domain blocks will be high enough to achieve significant compression. Along with this, the fractal encoding process can take quite some time while a search is performed for matching blocks.

CHAPTER 3
JPEG COMPRESSION

3.1 JPEG Image Compression
One of the most widely used and most popular image compression formats is the Joint Photographic Experts Group (JPEG) image compression standard. In order to meet the various needs of today's vast number of applications, the JPEG image compression standard has two basic compression methods, each with various modes of operation. A DCT-based method is available for lossy compression and a predictive method for lossless compression [14]. The JPEG compression standard also features a simpler lossy technique commonly referred to as the baseline method. Clearly the most widely implemented JPEG compression method to date, the baseline method, is sufficient for most applications. To be JPEG compatible, a system or product must include support for the baseline algorithm. The JPEG compression standard supports the following four modes of operation [14] listed below and illustrated in Figure 3.1:

- Sequential encoding: each image component is encoded in a single left-to-right, top-to-bottom scan;
- Progressive encoding: the image is encoded in multiple scans for applications in which transmission time is long, and the viewer prefers to watch the image build up in multiple coarse-to-clearpasses;Lossless encoding: the image is encoded to guarantee exact recovery of every source image sample value (even though the result is low compression compared to the lossy modes);
- Hierarchical encoding: the image is encoded at multiple resolutions so that lower-resolution versions may be accessed without first having to decompress the image at its full resolution.

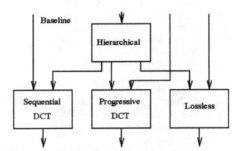

Figure 3.1 JPEG compression modes of operation [15].

3.2 JPEG baseline processing steps
A simplified block diagram showing the four sequential steps performed by the JPEG algorithm is shown in Figure 3.2. The first step is to take the source image and begin subdividing it into 8×8 nonoverlapping pixel blocks. These blocks are then sequentially encoded and processed from left to right, top to bottom through a Forward Discrete Cosine Transform (FDCT). As each 8x8 block is encoded, its 2-D DCT transform is computed. The next step is to then quantize, according to Eq. (3.1), and normalize the resulting coefficients, which will eliminate some of the less important information and improve data compression. After quantization, the DC coefficients are encoded using a Differential Pulse Code Modulation (DPCM) algorithm. The remaining AC coefficients are rearranged in a zigzag pattern, shown in Figure 3.3, in order to maximize the probability of similar size components being located near each other in the resulting data set. To achieve greater compression, the zigzag components are then entropy encoded which results in the final JPEG image compressed data.

6

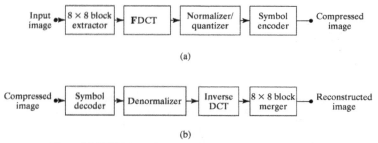

(a)

Compressed image → Symbol decoder → Denormalizer → Inverse DCT → 8 × 8 block merger → Reconstructed image

(b)

Figure 3.2 JPEG block diagram: (a) encoder and (b) decoder [11].

$$\hat{T}(u,v) = \text{round} \left[\frac{T(u,v)}{Z(u,v)} \right] \qquad (3.1)$$

where $\hat{T}(u,v)$ - Normalized and quantized coefficients,

$T(u,v)$ - 8×8 DCT sample of the image,

$Z(u,v)$ - Scaling transformation normalization array.

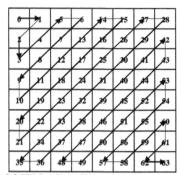

Figure 3.3 JPEG AC coefficient ordering zigzag pattern.

3.3 Discrete Cosine Transform (DCT) process

Figure 3.2a shows the key processing steps that are involved with the DCT-based modes of operation. Essentially DCT-based image compression can be thought of as the compression of a stream of grayscale image samples measuring 8x8 blocks in size. On the other hand, color image compression can be approximately treated as the compression of multiple grayscale images, either compressed in their entirety, individually one at a time, or compressed alternatively by interleaving 8x8 sample blocks from each in turn.

The image compression process begins with the source image being grouped into 8x8 pixel blocks, processed from left to right, top to bottom. Following this process, each 8x8 block in the image is then level shifted from unsigned integers having a range of $[0, 2^{n} - 1]$ to signed integers with a range of $[-2^{n-1}, 2^{n-1} - 1]$ by subtracting the quantity 2^{n-1}. Here 2^{n} represents the maximum number of gray levels in the image. The 2-D FDCT for each 8x8 pixel block is then computed according to Eq. (3.2) [1,14].

7

$$F(u,v) = \frac{1}{4} C(u) C(v) \left[\sum_{x=0}^{7} \sum_{y=0}^{7} f(x,y) * \cos\frac{(2x+1)}{16} \cos\frac{(2y+1)}{16} \right] \quad (3.2)$$

where: $C(u), C(v) = 1/\sqrt{2}$ for $u, v = 0$;

$C(u), C(v) = 1$ otherwise.

Each 8x8 pixel block sample from the source image can then be represented as a function of the two spatial dimensions x and y. The FDCT then takes these pixel values, decomposes them, and generates an output frequency whose values are the 64 DCT coefficient values. These values are considered to be the relative magnitudes of the 2-D spatial frequencies that exist in the 8x8 pixel block input signal. From this, the coefficient having a frequency coordinates of zero in both spatial dimensions is defined as the DC coefficient. The DC coefficient is simply a measure of the average value for the 64-pixel input signal. The remaining 63 coefficient values are called the AC coefficients. In addition, because the sample pixel values usually change very slowly from one point to another over an entire image, this processing step is essentially responsible for laying out the foundation to attain data compression. In order to accomplish this, most of the signal values are concentrated to the lower spatial frequencies since usually, but not always, the lower frequencies will have higher magnitudes. This step is extremely important in the process since for most samples taken from typical images, nearly all of the spatial frequencies will have a magnitude of either zero or near zero and thus do not need to be encoded.

3.4 Normalization/Quantization process

The normalization or quantization process of the JPEG image compression method is considered to be lossy. This comes from performing what is sometimes called a "many-to-one" mapping process in order to achieve good compression results by simply discarding any information considered not to be visually important. From the output of the FDCT process, every one of the 64 DCT coefficients is quantized uniformly in conjunction with a quantization table consisting of 64 elements. This quantization table must be supplied by either the user or the application will use a JPEG-defined default as the encoder input. Normalization or quantization is simply the process by which each DCT coefficient is divided by its corresponding quantizer step size, and the result rounded to the nearest integer (see Eq. 3.3).

$$F^Q(u,v) = IntegerRound\left(\frac{F(u,v)}{Q(u,v)}\right) \quad (3.3)$$

where: $F^Q(u,v)$ = the rounded and quantized DCT coefficient,

$F(u,v)$ = DCT sample (u,v) from the 8x8 image source block,

$Q(u,v)$ = the quantization table step size at (u,v).

After the quantization process is complete, the DC coefficient is isolated and treated separately from the remaining 63 AC coefficients. The DC coefficients of adjacent 8x8 pixel blocks have a strong correlation among each other. Due to this correlation, the quantized DC coefficients are encoded as the difference in the DC term from the previous 8x8 pixel block in the order of encoding. This helps to reduce the number of bits in the compressed image data. This encoding technique is known as Differential Pulse Code Modulation (DPCM) and is illustrated in Figure 3.4.

At this point all of the quantized coefficients, both AC and DC, are grouped and ordered into a zigzag type pattern as shown in Figure 3.3 in preparation for entropy coding. The ordering of the AC and DC coefficients in this manner aids in the entropy coding by placing all low frequency coefficients before the high frequency coefficients. The reason for doing this is that the low frequency coefficients are much more likely to be nonzero than the high frequency coefficients.

8

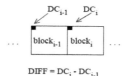

DIFF = DC_i - DC_{i-1}

Figure 3.4 Differential DC Encoding [14].

3.5 Entropy coding

The last step in this DCT-based encoder process is entropy coding. The purpose of this step is to hopefully achieve some additional lossless compression by encoding the already quantized DCT coefficients in a more compact manner based on their statistical characteristics. The JPEG image compression standard supports two entropy coding methods: Huffman [16] and arithmetic coding [17]. The JPEG baseline sequential DCT uses Huffman entropy coding simply to reduce the average number of bits relating a specific value to a unique code and is also intended for images having 8-bit samples. Another difference between the baseline coding method and the other sequential DCT coding methods is that its decoder only has the ability for storing two sets of Huffman tables. One table is for the AC coefficients and the other for the DC coefficients.

Entropy coding is better understood if viewed as a two step process. In the first step a conversion takes place between the zigzag sequence of the quantized AC and DC coefficients to some intermediate symbol sequence [14]. In the second step of this process the intermediate sequence of symbols created in the first step are then converted into a data stream. The entropy coding method used and the DCT-based operation mode chosen play an important role in the determination of the intermediate symbol set.

One of the requirements of Huffman coding is that at least one of the Huffman code table sets be specified by the source application. The reason for this is that the same tables used to compress the source image are also required in order to decompress it. The JPEG compression standard allows for the predefinition of Huffman tables and their use by the calling application as defaults, or the tables can be determined computationally for the given image in the first pass before compression. The JPEG standard does not require any Huffman tables to be specified. Instead this is left up to the calling application implementing JPEG compression.

CHAPTER 4
HYBRID ALGORITHM

4.1 Introduction
Over the past several years many hybrid image compression algorithms have been developed. Many researchers have developed hybrid algorithms which use only part of an image encoded using fractal methods and the remaining parts encoded using other methods. The sole purpose of the hybrid algorithm is to combine the advantages of two different coding algorithm methods to help improve the overall efficiency. Many attempts over the past several years have been made to try and combine fractal methods with other methods like vector quantization. Hamzaoui, Muller, and Saupe [18] developed a hybrid algorithm using vector quantization. Their scheme was to encode the source input image by combining both vector quantization codebook blocks and fractal coded image blocks together in hopes of achieving greater compression than if either method was used alone.

Other algorithms have been developed which combine fractal image compression methods hybridized with DCT image compression methods. Thao [19] came up with a fractal/DCT-based hybrid algorithm that used fractal encoding only on large contours and smooth variations within images and DCT encoding on the remaining parts of the image. The algorithm developed by Melnikov and Katsaggelos [20] used DCT encoding for the low frequency coefficients at predefined locations within an image block. The remaining coefficients were to be encoded using fractal methods. Curtis, Neil, and Fotopoulos [21] incorporated an FDCT pruning method developed by Skodras [22] in their hybrid algorithm. Their hopes were to increase the final image compression ratio while simultaneously reducing the overall DCT encoding time required.

The proposed hybrid algorithm introduced here combines the familiar DCT-based JPEG compression standard along with a searchless fractal encoding algorithm. The fractal algorithm is based on a quadtree partitioning of the images to be compressed and is capable of much deeper quadtrees than traditional algorithms, extending to single-pixel blocks [23]. The resulting hybrid compression algorithm yields greater compression ratios and improved image quality reconstruction than either JPEG or fractal compression methods alone.

4.2 Fractal method
The fractal image compression method used for the hybrid algorithm is a fast, searchless algorithm capable of very deep quadtree spanning and high compression ratios. Traditional methods try to identify self-similarity among the image elements via an exhaustive search of the pixel domain space of the image leading to an $O(n^4)$ algorithm [23]. This high order algorithm gives rise to very slow execution times, especially as the image sizes increase. The searchless fractal algorithm described here avoids this process and yields a simpler $O(n^2)$ algorithm and much faster execution times. The fractal portion of the hybrid algorithm uses an user-specified MSE tolerance value to determine which image elements exhibit acceptable self-similarity. Additionally, this error tolerance value is used to determine if an image block is of very low contrast so that it may be encoded as a simple gray block [25, 26].

The fractal portion of the hybrid algorithm begins by partitioning the original image up into a set of non-overlapping ranges, which when taken as a set, comprise the pixels of the original image. The largest ranges considered are 32x32 pixels and the quadtree incorporates all smaller sub-ranges down to the single-pixel level (*i.e.* 1x1 block) [23]. This is a great improvement over traditional fractal methods which typically do not support block sizes smaller than 4x4 pixels. For every *NxN* pixel range block, the best self-similarity parameters are determined by the algorithm for each *2Nx2N* domain having a center point at the same locus as the center point of the range. For those ranges close enough to the edges of the image that the domain extends beyond the boundaries of the image, the closest domain within the boundaries of the image is chosen. There is some loss of image quality introduced by eschewing the domain search process but this loss is more than offset by the greater fidelity allowed by the support of deeper quadtrees [23, 26, 27].

10

4.3 DCT-based JPEG method

Once the fractal data for the image has been calculated and the fractal approximation for the image has been reconstructed, the hybridization process proceeds on to the JPEG algorithm. The original image is then compressed and reconstructed using the JPEG algorithm, and a user-specified tolerance level ranging from 0 to 100. The higher the number, the better the quality (less image degradation due to compression), but the resulting reconstructed file will also be larger. The JPEG algorithm is very flexible, so it can create relatively large files with excellent image quality or very small files with relatively poor image quality. The user can define the amount of compression desired dependent on the needs of the application [28].

The DCT-based JPEG algorithm encodes the low-contrast blocks exhibiting smooth transitions very efficiently while the fractal method exhibits greater efficiency for the high-contrast blocks. Within the JPEG algorithm, the quantized-DCT process always decomposes an image into 8x8 pixel blocks. In order to obtain the frequency-domain components, the DCT is applied to each pixel block. The resulting frequency-domain components are then quantized with the level of quantization controlling the quality and CR of this process [23]. Following this step, the quantized components are then run-length encoded and losslessly compressed.

4.4 Hybridization method

After the entire image has been compressed using both fractal and JPEG methods, the hybridization process begins. Starting at the upper-left corner, the quality of the compressed image from both the fractal and JPEG method is compared and whichever yields a higher PSNR for that block is retained. When comparing blocks of the image to comprise the hybridized image, sufficient blocks produced by each method are used to make the total areas equal [23]. As an example, a single $16x16$ fractal block from the quadtree would be compared to four 8x8 JPEG blocks or four $4x4$ fractal blocks would be compared to a single 8x8 JPEG block.

As a result of this comparison process, a mosaic emerges in which some portions of the original image are represented using DCT-based JPEG data and others are represented using fractal data. In order to identify which blocks within the resulting image are encoded as fractal data and which are encoded as JPEG DCT-based data additional overhead is required for the hybrid image data. In this hybrid image algorithm implementation, the amount of additional overhead has been streamlined and losslessly compressed so as to minimize the impact on the overall size of the hybrid image. Additionally, a non-traditional form for the fractal transform of each range block is implemented as shown in Eq. (4.1) [25, 26, 27, 29].

$$W(D) = s\left(D - \bar{d}I\right) + \bar{r}I \qquad (4.1)$$

In Equation (4.1), D represents the domain data, $W(D)$ represents the transformed domain, s represents the contrast scale factor, I represents the identity matrix, and $\bar{d}$ and $\bar{r}$ are the mean gray values for the domain and range, respectively. By using this transform the quantity of data stored for each range is reduced to only the scale factor s and the range average $\bar{r}$. Additionally, the implementation of variable-length encoding for the fractal data helps to improve the overall compression ratio of the hybridized image.

CHAPTER 5
2-WAY HYBRID TEST IMAGE COMPARISONS

5.1 Introduction
 The test images chosen for the hybrid algorithm are listed in Table 5.1. These images were chosen for their combination of various complexities and textures containing smooth and contoured areas. For comparison of the final results from the fractal, JPEG, and hybrid algorithms two measurements were used. These measurements are the compression ratio (CR) and PSNR. The compression ratio measures the compactness of the resulting image with respect to the original image according to Eq. (5.1).

$$C_R = \frac{S_1}{S_2},$$ (5.1)

 where S_1 = size of the original image

  S_2 = size of the compressed image

The PSNR measures the image distortion of the reconstructed image $I'(x, y)$ from the original image $I(x, y)$. The PSNR is parameterized with the MSE which is the cumulative mean squared error between the original image and the reconstructed image. In Eq. (5.2) M, N are the dimensions of the image.

$$MSE = \frac{1}{MN} \sum_{y=1}^{M} \sum_{x=1}^{N} \left[I(x, y) - I'(x, y) \right]^2$$ (5.2)

$$PSNR = 20 \cdot \log_{10} \frac{255}{MSE^{1/2}}$$ (5.3)

Table 5.1 Image dimension sizes in pixels for test images used

Image Name	Image Size				
	64×64	128×128	256×256	512×512	768×768
Boy		X	X		
Crowd		X	X		
Fruit	X	X	X	X	
Kameraman		X	X	X	
Lena		X	X	X	
Mosaic		X	X	X	X
Orchid		X	X	X	
Peacock		X	X	X	
Peppers		X	X	X	
Waterfall		X	X	X	

5.2 Results for Boy128.pgm Image

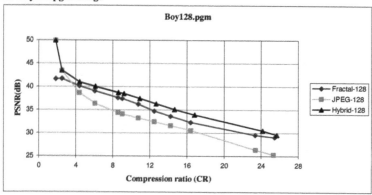

Figure 5.1 Fractal, JPEG, and hybrid image compression results for Boy128.pgm.

The graph shown in Figure 5.1 shows the improvement in CR when using the hybrid algorithm versus the fractal or JPEG compression algorithm alone. The image itself is an 8-bit grayscale image measuring 128x128 pixels as shown in part (a) of Figure 5.2. Using just the fractal algorithm, the compression ratio ranges from 1.844 to 25.464, the PSNR from 41.664 down to 29.069, with the MSE being varied from 1.6 to 16.8. If just the JPEG compression algorithm is used, the compression ratio varies from 1.84 to 25.346 and the PSNR from 49.976 to 25.288 for quality levels of 98 down to 4. When used individually, both fractal and JPEG compression algorithms yield similar results, with the fractal algorithm producing slightly better results. Compressed using the hybrid algorithm, the compression ratio varies from 1.836 to 25.704 and the PSNR from 49.976 to 29.604. Overall, the gain in compression ratio using the hybrid algorithm is approximately 1.0%. The chart shown in Figure 5.1 was generated from the data listed in Table 5.2. Figure 5.2 shows a comparison of the compressed image results using the hybrid, JPEG, and fractal algorithms. Part (a) of Figure 5.2 is the original image. Part (b) of Figure 5.2 is the image compressed by the 2-way hybrid with CR=25.704 and PSNR=29.604. Part (c) of Figure 5.2 is the image compressed by JPEG with Quality=4, CR=25.346 and PSNR=25.288. Part (c) exhibits the loss of detail typical of JPEG at high CR [31]. Part (d) of Figure 5.2 is the image compressed by fractal with MSE=16.8, CR=25.464 and PSNR=29.069. Part (d) exhibits the "blockiness" typical of fractal methods at high CR [31]. Clearly part (b) is a much higher quality image having not only a better compression ratio but also a higher PSNR than either part (c) or part (d).

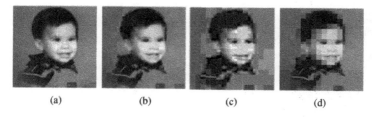

| (a) | (b) | (c) | (d) |

Figure 5.2 Boy128.pgm image at CR approximately 26:1.

Table 5.2 Compression algorithm results for Boy128.pgm image file

Fractal			JPEG			Hybrid	
MSE	CR	PSNR	Quality	CR	PSNR	CR	PSNR
1.6	1.844	41.664	98	1.840	49.976	1.836	49.976
2.7	2.504	41.706	95	2.503	43.418	2.506	43.467
3.9	4.397	40.134	87	4.391	38.677	4.414	40.964
4.5	6.051	39.091	77	6.071	36.356	6.124	40.035
5.7	8.563	37.631	58	8.568	34.392	8.613	38.744
6	9.040	37.412	53	9.030	34.081	9.203	38.484
7.2	10.746	36.238	38	10.739	33.210	10.904	37.452
8.7	12.471	34.762	28	12.471	32.423	12.673	36.305
10	14.248	33.589	21	14.248	31.557	14.668	35.097
11.2	16.415	32.317	15	16.415	30.537	16.819	33.949
16	23.360	29.548	5	23.394	26.343	24.152	30.508
16.8	25.464	29.069	4	25.346	25.288	25.704	29.604

5.3 Results for Boy256.pgm Image

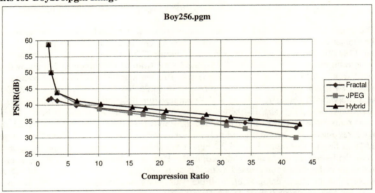

Figure 5.3 Fractal, JPEG, and hybrid image compression results for Boy256.pgm.

The graph shown in Figure 5.3 shows the improvement in CR when using the hybrid algorithm versus the fractal or JPEG compression algorithm alone. The image itself is an 8-bit grayscale image measuring 256x256 pixels as shown in part (a) of Figure 5.4. Using just the fractal algorithm, the compression ratio ranges from 1.791 to 42.346, the PSNR from 41.551 down to 32.657, with the MSE being varied from 1.2 to 11.1. The JPEG compression algorithm yields compression ratios varying from 1.784 to 42.346 and PSNR values from 58.741 to 29.637 for quality levels of 100 down to 6. Used individually, both fractal and JPEG compression algorithms result in almost identical results. Compressed using the hybrid algorithm, the compression ratio varies from 1.78 to 42.956 and the PSNR from 58.741 to 33.8. The chart shown in Figure 5.3 was generated from the data listed in Table 5.3. Shown in Figure 5.4 are the resulting reconstructed images using the hybrid, JPEG, and fractal image compression algorithms. Part (a) of Figure 5.4 is the original image. Part (b) of Figure 5.4 is the image compressed by the 2-way hybrid with CR=42.956 and PSNR=33.8. Part (c) of Figure 5.4 is the image compressed by JPEG with Quality=6, CR=42.346 and PSNR=29.637. Part (c) exhibits the loss of detail typical of JPEG at high CR [31]. Part (d) of Figure 5.4 is the image compressed by fractal with MSE=11.1, CR=42.346 and PSNR=32.657. Part (d) exhibits excessive blockiness typical of fractal

14

methods. Clearly part (b) is a much higher quality image having not only a better compression ratio but also a higher PSNR than either part (c) or part (d).

Table 5.3 Compression algorithm results for Boy256.pgm image file

Fractal			JPEG			Hybrid	
MSE	CR	PSNR	Quality	CR	PSNR	CR	PSNR
1.2	1.791	41.551	100	1.784	58.741	1.780	58.741
2.2	2.213	42.017	98	2.217	50.058	2.210	50.058
3	3.233	41.349	95	3.206	43.779	3.199	43.793
4	6.373	39.916	87	6.382	40.494	6.483	41.257
4.5	10.128	39.062	74	10.130	38.794	10.361	40.298
5.1	15.188	38.108	55	15.195	37.488	15.541	39.360
5.5	17.342	37.711	44	17.342	36.981	17.673	38.962
6.2	20.659	36.938	32	20.639	36.141	21.071	38.251
7.9	27.065	35.626	19	27.065	34.535	27.682	36.934
8.8	30.949	34.677	14	30.949	33.449	31.667	36.082
9.5	33.999	34.261	11	34.052	32.469	34.886	35.495
11.1	42.346	32.657	6	42.346	29.637	42.956	33.800

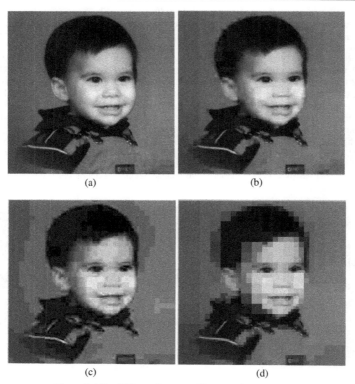

(a) (b)

(c) (d)

Figure 5.4 Boy256.pgm image at CR approximately 42:1.

15

5.4 Results for Crowd128.pgm Image

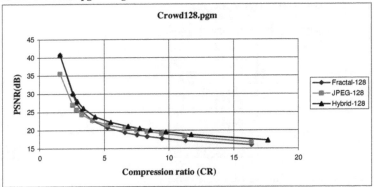

Figure 5.5 Fractal, JPEG, and hybrid image compression results for Crowd128.pgm.

     The graph shown in Figure 5.5 shows a slight improvement by using the hybrid algorithm over the fractal and JPEG methods. The original image file, shown in part (a) of Figure 5.6, is an 8-bit grayscale image measuring 128x128 pixels. Using only the fractal algorithm, the compression ratio ranges from 1.604 to 16.383, the PSNR from 40.807 down to 16.029, with the MSE being varied from 1.0 to 49.6. The JPEG compression algorithm yields compression ratios varying from 1.603 to 16.366 and PSNR values from 35.538 to 16.678 for quality levels of 89 down to 3. Used individually, both fractal and JPEG compression algorithms produce almost identical results. Compressed using the hybrid algorithm, the compression ratio varies from 1.6 to 17.652 and the PSNR from 40.824 to 17.235. The chart shown in Figure 5.5 was generated from the data listed in Table 5.4. Shown in Figure 5.6 are the resulting reconstructed images using the hybrid, JPEG, and fractal image compression algorithms. Part (a) of Figure 5.6 is the original image. Part (b) of Figure 5.6 is the image compressed by the 2-way hybrid with CR=5.496 and PSNR=22.247. Part (c) of Figure 5.4 is the image compressed by JPEG with Quality=22, CR=5.263 and PSNR=21.531. Part (d) of Figure 5.6 is the image compressed by fractal with MSE=32.3, CR=5.253 and PSNR=20.767. Part (d) exhibits excessive blockiness typical of fractal methods. Clearly part (b) is a much higher quality image having not only a better compression ratio but also a higher PSNR than either part (c) or part (d).

(a)   (b)   (c)   (d)

Figure 5.6 Crowd128.pgm image at CR approximately 5:1.

Table 5.4 Compression algorithm results for Crowd128.pgm image file

Fractal			JPEG			Hybrid	
MSE	CR	PSNR	Quality	CR	PSNR	CR	PSNR
1.0	1.604	40.807	89	1.603	35.538	1.600	40.824
13.9	2.564	29.967	67	2.563	26.989	2.596	30.159
17.2	2.865	27.569	59	2.864	25.676	2.924	27.966
21.1	3.265	25.434	48	3.269	24.395	3.390	26.072
26.9	4.096	22.720	33	4.101	22.823	4.290	23.754
32.3	5.253	20.767	22	5.263	21.531	5.496	22.247
37.0	6.612	19.432	15	6.615	20.560	6.847	21.156
39.3	7.543	18.747	12	7.533	20.066	7.743	20.599
40.7	8.320	18.305	10	8.324	19.660	8.555	20.100
42.8	9.468	17.811	8	9.441	19.126	9.756	19.554
45.3	11.286	17.137	6	11.232	18.386	11.705	18.805
49.6	16.383	16.029	3	16.366	16.678	17.652	17.235

5.5 Results for Crowd256.pgm Image

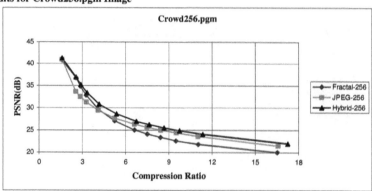

Figure 5.7 Fractal, JPEG, and hybrid image compression results for Crowd256.pgm.

The graph shown in Figure 5.7 shows a slight improvement by using the hybrid algorithm over the fractal and JPEG methods. The original image file, shown in Figure 5.8, is an 8-bit grayscale image measuring 256x256 pixels. Using only the fractal algorithm, the compression ratio ranges from 1.601 to 16.52, the PSNR from 40.769 down to 20.051, with the MSE being varied from 0.1 to 34.8. The JPEG compression algorithm yields compression ratios varying from 1.594 to 16.578 and PSNR values from 40.944 to 21.53 for quality levels of 94 down to 6. Used individually, both fractal and JPEG compression algorithms produce almost identical results. Compressed using the hybrid algorithm, the compression ratio varies from 1.59 to 17.241 and the PSNR ranges from 41.243 to 22.018. The chart shown in Figure 5.7 was generated from the data listed in Table 5.5. Shown in Figure 5.8 are the resulting reconstructed images using the hybrid, JPEG, and fractal image compression algorithms. Part (a) of Figure 5.8 is the original image. Part (b) of Figure 5.8 is the image compressed by the 2-way hybrid with CR=17.241 and PSNR=22.018. Part (c) of Figure 5.8 is the image compressed by JPEG with Quality=6, CR=16.578 and PSNR=21.53. Part (d) of Figure 5.8 is the image compressed by fractal with MSE=34.8, CR=16.52 and PSNR=20.051. Part (d) exhibits excessive blockiness typical of fractal methods. Clearly part (b) is a

17

much higher quality image having not only a better compression ratio but also a higher PSNR than either part (c) or part (d).

Table 5.5 Compression algorithm results for Crowd256.pgm image file

Fractal			JPEG			Hybrid	
MSE	CR	PSNR	Quality	CR	PSNR	CR	PSNR
0.1	1.601	40.769	94	1.594	40.944	1.590	41.243
6.3	2.564	36.762	83	2.553	33.663	2.566	36.964
7.6	2.864	34.821	79	2.861	32.477	2.894	35.234
9.3	3.265	32.783	73	3.270	31.223	3.318	33.436
12.8	4.094	29.671	59	4.080	29.401	4.165	30.765
17.1	5.244	27.052	39	5.261	27.668	5.380	28.618
21.0	6.619	25.048	26	6.620	26.313	6.759	26.980
23.0	7.509	24.132	21	7.489	25.647	7.636	26.214
24.9	8.426	23.401	17	8.438	24.975	8.660	25.489
26.9	9.508	22.630	14	9.535	24.373	9.758	24.851
29.1	11.050	21.823	11	11.032	23.612	11.349	24.078
34.8	16.520	20.051	6	16.578	21.530	17.241	22.018

(a)   (b)

(c)   (d)

Figure 5.8 Crowd256.pgm image at CR approximately 17:1.

5.6 Results for Fruit64.pgm Image

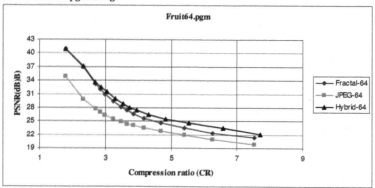

Figure 5.9 Fractal, JPEG, and hybrid image compression results for Fruit64.pgm.

The graph shown in Figure 5.9 shows an improvement by using the hybrid algorithm over the fractal and JPEG methods. The original image file, shown in part (a) of Figure 5.10, is an 8-bit grayscale image measuring 64x64 pixels. For the fractal algorithm, the compression ratio ranges from 1.794 to 7.539, the PSNR from 40.895 down to 21.348, with an MSE from 3.17 to 31.83. The JPEG compression algorithm shows compression ratios varying from 1.794 to 7.539 and PSNR values from 34.864 to 19.870 for quality levels of 87 to 6. Used individually, both fractal and JPEG compression algorithms produce similar results, but for this image the fractal algorithm produces better results. Compressed using the hybrid algorithm compression ratios vary from 1.790 to 7.709 with PSNR values from 40.895 to 22.120. The chart shown in Figure 5.9 was generated from the data listed in Table 5.6. Shown in Figure 5.10 are the resulting reconstructed images using the hybrid, JPEG, and fractal image compression algorithms. Part (a) of Figure 5.10 is the original image. Part (b) of Figure 5.10 is the image compressed by the 2-way hybrid with CR=3.3 and PSNR=29.883. Part (c) of Figure 5.10 is the image compressed by JPEG with Quality=47, CR=3.248 and PSNR=25.423. Part (d) of Figure 5.10 is the image compressed by fractal with MSE=14.74, CR=3.248 and PSNR=29.432. Part (d) exhibits excessive blockiness typical of fractal methods. Clearly part (b) is a much higher quality image having not only a better compression ratio but also a higher PSNR than either part (c) or part (d).

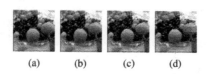

| (a) | (b) | (c) | (d) |

Figure 5.10 Fruit64.pgm image at CR approximately 3:1.

Table 5.6 Compression algorithm results for Fruit64.pgm image file

Fractal			JPEG			Hybrid	
MSE	CR	PSNR	Quality	CR	PSNR	CR	PSNR
3.17	1.794	40.895	87	1.794	34.864	1.790	40.895
6.93	2.321	36.969	75	2.321	29.791	2.336	37.049
10.44	2.698	33.168	65	2.698	27.678	2.691	33.616
11.47	2.840	32.090	60	2.840	26.939	2.861	32.530
12.66	2.984	30.929	55	2.984	26.273	3.053	31.465
14.74	3.248	29.432	47	3.248	25.423	3.300	29.883
16.41	3.482	28.118	41	3.482	24.840	3.542	28.799
17.83	3.652	27.409	37	3.652	24.395	3.735	28.015
18.99	3.865	26.641	33	3.865	23.994	3.951	27.361
21.00	4.180	25.625	27	4.180	23.428	4.316	26.484
23.36	4.685	24.651	21	4.685	22.727	4.840	25.496
25.73	5.407	23.449	15	5.407	21.948	5.553	24.562
28.60	6.254	22.344	10	6.254	21.033	6.574	23.435
31.83	7.539	21.348	6	7.539	19.870	7.709	22.120

5.7 Results for Fruit128.pgm Image

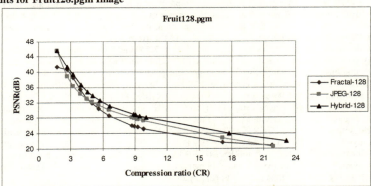

Figure 5.11 Fractal, JPEG, and hybrid image compression results for Fruit128.pgm.

The graph shown in Figure 5.11 shows an improvement by using the hybrid algorithm over the fractal and JPEG methods. The original image file, shown in part (a) of Figure 5.12, is an 8-bit grayscale image measuring 128x128 pixels. For the fractal algorithm, the compression ratio ranges from 1.731 to 21.807, the PSNR from 41.201 down to 20.749, with an MSE from 1.2 to 32.7. The JPEG compression algorithm shows compression ratios varying from 1.731 to 21.953 and PSNR values from 45.524 to 20.567 for quality levels of 96.8 to 3. Used individually, both fractal and JPEG compression algorithms produce fairly similar results, with JPEG yielding slightly higher results. Compressed using the hybrid algorithm compression ratios vary from 1.73 to 23.13 with PSNR values from 45.524 to 21.996. The chart shown in Figure 5.11 was generated from the data listed in Table 5.7. Shown in Figure 5.12 are the resulting reconstructed images using the hybrid, JPEG, and fractal image compression algorithms. Part (a) of Figure 5.12 is the original image. Part (b) of Figure 5.12 is the image compressed by the 2-way hybrid with CR=9.441 and PSNR=28.469. Part (c) of Figure 5.12 is the image compressed by JPEG with Quality=17.8, CR=9.249 and PSNR=27.633. Part (d) of Figure 5.12 is the image compressed by fractal with MSE=22, CR=9.249 and PSNR=25.526. Part (d) exhibits excessive blockiness typical of fractal

methods. Clearly part (b) is a much higher quality image having not only a better compression ratio but also a higher PSNR than either part (c) or part (d).

| (a) | (b) | (c) | (d) |

Figure 5.12 Fruit128.pgm image at CR approximately 9:1.

Table 5.7 Compression algorithm results for Fruit128.pgm image file

Fractal			JPEG			Hybrid	
MSE	CR	PSNR	Quality	CR	PSNR	CR	PSNR
1.2	1.731	41.201	96.8	1.731	45.524	1.730	45.524
4	2.679	40.547	89.8	2.673	38.718	2.662	41.240
5.2	3.229	38.379	83.8	3.229	36.287	3.241	39.341
7.2	3.911	35.511	75.8	3.911	34.303	3.951	36.694
9.1	4.493	33.082	66.8	4.493	32.941	4.564	34.843
10.7	4.981	31.822	58.8	4.980	32.132	5.063	33.754
12.7	5.606	30.317	47.8	5.605	31.294	5.702	32.568
15.3	6.552	28.615	34.8	6.552	30.097	6.653	31.085
20.9	8.737	25.909	19.8	8.737	28.024	8.888	28.920
21.4	8.947	25.714	18.8	8.947	27.836	9.105	28.740
22	9.249	25.526	17.8	9.249	27.633	9.441	28.469
22.8	9.826	25.146	15.8	9.820	27.203	10.042	28.076
29.4	17.172	21.635	5.8	17.154	22.738	17.786	23.875
32.7	21.807	20.749	3	21.953	20.567	23.130	21.996

5.8 Results for Fruit256.pgm Image

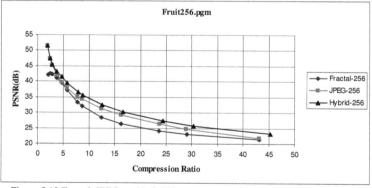

Figure 5.13 Fractal, JPEG, and hybrid image compression results for Fruit256.pgm.

21

The graph shown in Figure 5.13 shows an improvement by using the hybrid algorithm over the fractal and JPEG methods. The original image file, shown in part (a) of Figure 5.14, is an 8-bit grayscale image measuring 256x256 pixels. For the fractal algorithm, the compression ratio ranges from 1.972 to 43.097, the PSNR from 42.053 down to 21.508, with an MSE from 1.3 to 31.6. The JPEG compression algorithm shows compression ratios varying from 1.967 to 43.154 and PSNR values from 51.507 to 21.757 for quality levels of 98.8 to 2.8. Used individually, both fractal and JPEG compression algorithms produce fairly similar results, with JPEG producing better results once again. Compressed using the hybrid algorithm compression ratios vary from 1.963 to 45.27 with PSNR values from 51.507 to 23.253. The chart shown in Figure 5.13 was generated from the data listed in Table 5.8. Shown in Figure 5.14 are the resulting reconstructed images using the hybrid, JPEG, and fractal image compression algorithms. Part (a) of Figure 5.14 is the original image. Part (b) of Figure 5.14 is the image compressed by the 2-way hybrid with CR=8.737 and PSNR=35.641. Part (c) of Figure 5.14 is the image compressed by JPEG with Quality=41.8, CR=8.605 and PSNR=34.245. Part (d) of Figure 5.14 is the image compressed by fractal with MSE=10.2, CR=8.605 and PSNR=32.136. Part (d) exhibits excessive blockiness typical of fractal methods. Clearly part (b) is a much higher quality image having not only a better compression ratio but also a higher PSNR than either part (c) or part (d).

Table 5.8 Compression algorithm results for Fruit256.pgm image file

Fractal			JPEG			Hybrid	
MSE	CR	PSNR	Quality	CR	PSNR	CR	PSNR
1.3	1.972	42.053	98.8	1.967	51.507	1.963	51.507
2.2	2.364	42.542	96.8	2.364	47.507	2.364	47.508
3	2.850	42.336	94.8	2.820	45.185	2.815	45.289
3.8	3.720	41.168	89.8	3.745	41.944	3.733	43.209
4.7	4.705	39.398	82.8	4.703	39.618	4.730	41.443
6	5.700	37.142	73.8	5.710	37.768	5.758	39.449
9.1	7.801	33.284	49.8	7.803	35.005	7.917	36.485
10.2	8.605	32.136	41.8	8.605	34.245	8.737	35.641
15.8	12.431	28.485	21.8	12.429	31.264	12.616	32.429
19.9	16.270	26.300	13.8	16.274	29.202	16.595	30.165
25	23.665	24.052	7.8	23.673	26.401	24.350	27.279
27.4	29.031	23.049	5.8	28.979	24.753	30.376	25.782
31.6	43.097	21.508	2.8	43.154	21.757	45.270	23.253

(a)   (b)

(c)   (d)

Figure 5.14 Fruit256.pgm image at CR approximately 9:1.

5.9 Results for Fruit512.pgm Image

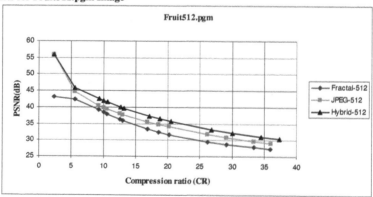

Figure 5.15 Fractal, JPEG, and hybrid image compression results for Fruit512.pgm.

The graph shown in Figure 5.15 shows an improvement by using the hybrid algorithm over the fractal and JPEG methods. The original image file, shown in Figure 5.16, is an 8-bit grayscale image measuring 512x512 pixels. For the fractal algorithm, the compression ratio ranges from 2.305 to 35.912, the PSNR from 43.047 down to 27.42, with an MSE from 1.2 to 17.4. The JPEG compression algorithm shows compression ratios varying from 2.314 to 35.991 and PSNR values from 55.971 to 29.039 for quality levels of 99 to 7. Used individually, both fractal and JPEG compression algorithms produce fairly similar results, with JPEG producing better results once again. Compressed using the hybrid algorithm compression ratios vary from 2.285 to 37.35 with PSNR values from 55.971 to 30.471. The chart shown in Figure 5.15 was generated from the data listed in Table 5.9. For this comparison, the image samples were all selected at the data points where CR is approximately 13:1. Shown in Figure 5.17 is the image compressed by the 2-way hybrid with CR=13.15 and PSNR=39.541. Figure 5.18 is the image compressed by JPEG with Quality=41, CR=13.026, and PSNR=37.649. Figure 5.19 is the image compressed by fractal with MSE=7, CR=13.036, and PSNR=35.831. The 2-way hybrid eliminates the worst of the problems seen in fractal or JPEG alone [31]. Subjectively, the 2-way hybrid in Figure 5.17 is better than the other methods.

Table 5.9 Compression algorithm results for Fruit512.pgm image file

Fractal			JPEG			Hybrid	
MSE	CR	PSNR	Quality	CR	PSNR	CR	PSNR
1.2	2.305	43.047	99	2.314	55.971	2.285	55.971
3.6	5.612	42.231	88	5.624	44.648	5.626	45.830
4.9	9.253	39.147	68	9.262	40.389	9.299	42.567
5.3	10.027	38.452	63	10.011	39.796	10.073	41.890
5.6	10.570	37.899	59	10.577	39.392	10.645	41.437
6.7	12.545	36.197	44	12.543	37.965	12.655	39.880
7	13.036	35.831	41	13.026	37.649	13.150	39.541
9.2	16.854	33.430	26	16.880	35.537	17.121	37.249
10.1	18.545	32.411	22	18.548	34.780	18.826	36.393
11	20.177	31.600	19	20.177	34.050	20.518	35.607
14	26.242	29.384	12	26.219	31.808	26.841	33.160
15.3	29.142	28.690	10	29.171	30.867	30.006	32.228
16.7	33.451	27.841	8	33.434	29.724	34.517	31.092
17.4	35.912	27.420	7	35.991	29.039	37.350	30.471

Figure 5.16 Original Fruit512.pgm image.

Figure 5.17 Fruit512.pgm image using 2-way hybrid $(CR = 13.15, PSNR = 39.541)$.

Figure 5.18 Fruit512.pgm image using JPEG $(Quality = 41, CR = 13.026, PSNR = 37.649)$.

Figure 5.19 Fruit512.pgm image using fractal $(MSE = 7, CR = 13.036, PSNR = 35.831)$.

5.10 Results for Kameraman128.pgm Image

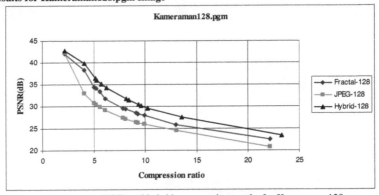

Figure 5.20 Fractal, JPEG, and hybrid compression results for Kameraman128.pgm.

The graph shown in Figure 5.20 shows an improvement by using the hybrid algorithm over the fractal and JPEG methods. The original image file, shown in part (a) of Figure 5.21, is an 8-bit grayscale image measuring 128x128 pixels. For the fractal algorithm, the compression ratio ranges from 2.191 to

28

22.131, the PSNR from 42.08 down to 22.247, with an MSE from 2.1 to 38.4. The JPEG compression algorithm shows compression ratios varying from 2.192 to 22.131 and PSNR values from 42.265 to 20.673 for quality levels of 94 to 3. Used individually, both fractal and JPEG compression algorithms produce similar results, with the fractal compression algorithm producing the better results. Compressed using the hybrid algorithm compression ratios vary from 2.182 to 23.261 with PSNR values from 42.816 to 23.345. The chart shown in Figure 5.20 was generated from the data listed in Table 5.10. Figure 5.21 shows examples of the Kameraman image compressed by each method. For this comparison, the image samples were all selected at the data points in Figure 5.20 where CR is approximately 6:1. Part (a) of Figure 5.21 is the original image. Part (b) of Figure 5.21 is the image compressed by the 2-way hybrid with CR=6.235 and PSNR=34.178. Part (c) of Figure 5.21 is the image compressed by JPEG with Quality=53, CR=6.11 and PSNR=29.197. Part (d) of Figure 5.21 is the image compressed by fractal with MSE=12, CR=6.112 and PSNR=31.779. Part (d) exhibits excessive blockiness typical of fractal methods. Clearly part (b) is a much higher quality image having not only a better compression ratio but also a higher PSNR than either part (c) or part (d).

| (a) | (b) | (c) | (d) |

Figure 5.21 Kameraman128.pgm image at CR approximately 6:1.

Table 5.10 Compression algorithm results for Kameraman128.pgm image file

Fractal			JPEG			Hybrid	
MSE	CR	PSNR	Quality	CR	PSNR	CR	PSNR
2.1	2.191	42.080	94	2.192	42.265	2.182	42.816
5.8	4.082	38.265	77	4.080	32.991	4.086	39.905
8.9	5.054	34.449	66	5.054	30.785	5.131	36.394
9.4	5.238	34.114	64	5.236	30.502	5.307	35.971
11	5.606	33.442	59	5.606	29.834	5.730	35.073
12	6.112	31.779	53	6.110	29.197	6.235	34.178
16	7.831	29.599	34	7.831	27.425	8.130	31.727
16	8.055	29.421	32	8.055	27.181	8.367	31.430
18	9.065	28.535	26	9.065	26.457	9.366	30.422
19	9.296	28.292	25	9.296	26.322	9.584	30.227
20	9.933	27.922	22	9.933	25.944	10.217	29.654
26	13.005	25.855	13	13.015	24.485	13.519	27.548
38	22.131	22.427	3	22.131	20.673	23.261	23.345

5.11 Results for Kameraman256.pgm Image

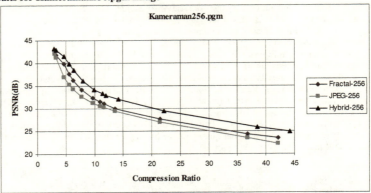

Figure 5.22 Fractal, JPEG, and hybrid compression results for Kameraman256.pgm.

The graph shown in Figure 5.22 shows an improvement by using the hybrid algorithm over the fractal and JPEG methods. The original image file, shown in part (a) of Figure 5.23, is an 8-bit grayscale image measuring 256x256 pixels. For the fractal algorithm, the compression ratio ranges from 3.03 to 42.128, the PSNR from 42.089 down to 23.497, with an MSE from 2.8 to 36. The JPEG compression algorithm shows compression ratios varying from 3.03 to 42.128 and PSNR values from 42.055 to 22.256 for quality levels of 93 to 2. Used individually, both fractal and JPEG compression algorithms produce similar results, with the fractal compression algorithm producing the better results. Compressed using the hybrid algorithm compression ratios vary from 3.022 to 44.231 with PSNR values from 43.15 to 24.925. The chart shown in Figure 5.22 was generated from the data listed in Table 5.11. Figure 5.23 shows examples of the Kameraman image compressed by each method. For this comparison, the image samples were all selected at the data points in Figure 5.22 where CR is approximately 8:1. Part (a) of Figure 5.23 is the original image. Part (b) of Figure 5.23 is the image compressed by the 2-way hybrid with CR=7.985 and PSNR=36.068. Part (c) of Figure 5.23 is the image compressed by JPEG with Quality=61, CR=7.821 and PSNR=32.65. Part (d) of Figure 5.23 is the image compressed by fractal with MSE=9.5, CR=7.821 and PSNR=34.151. Part (d) exhibits excessive blockiness typical of fractal methods. Clearly part (b) is a much higher quality image having not only a better compression ratio but also a higher PSNR than either part (c) or part (d).

30

Table 5.11 Compression algorithm results for Kameraman256.pgm image file

Fractal			JPEG			Hybrid	
MSE	CR	PSNR	Quality	CR	PSNR	CR	PSNR
2.8	3.030	42.089	93	3.030	42.055	3.022	43.150
3.1	3.270	41.792	92	3.276	41.253	3.268	42.927
4.5	4.627	39.913	84	4.631	37.004	4.653	41.526
5.9	5.551	37.659	78	5.560	35.341	5.605	39.917
7.1	6.279	36.222	73	6.285	34.346	6.358	38.451
9.5	7.821	34.151	61	7.821	32.650	7.985	36.068
12	9.727	32.320	45	9.736	31.221	9.989	34.157
13	11.019	31.523	37	11.017	30.532	11.408	33.287
14	11.654	31.070	34	11.654	30.275	11.992	32.943
16	13.631	29.964	26	13.620	29.407	14.112	32.015
23	21.422	27.580	12	21.429	27.011	22.198	29.394
33	36.744	24.277	4	36.744	23.539	38.514	25.770
36	42.128	23.497	2	42.128	22.256	44.231	24.925

(a)

(b)

(c)

(d)

Figure 5.23 Kameraman256.pgm image at CR approximately 8:1.

5.12 Results for Kameraman512.pgm Image

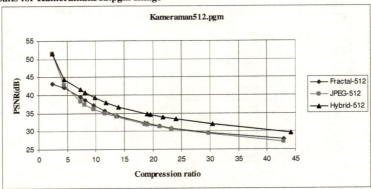

Figure 5.24 Fractal, JPEG, and hybrid compression results for Kameraman512.pgm.

The graph shown in Figure 5.24 shows an improvement by using the hybrid algorithm over the fractal and JPEG methods. The original image file, shown in Figure 5.25, is an 8-bit grayscale image measuring 512x512 pixels. For the fractal algorithm, the compression ratio ranges from 2.356 to 42.94, the PSNR from 43.1 down to 27.736, with an MSE from 1 to 23. The JPEG compression algorithm shows compression ratios varying from 2.358 to 42.963 and PSNR values from 51.543 to 27.159 for quality levels of 98 to 6. Used individually, both fractal and JPEG compression algorithms produce similar results, with the fractal compression algorithm producing the better results. Compressed using the hybrid algorithm compression ratios vary from 2.352 to 44.216 with PSNR values from 51.544 to 29.479. The chart shown in Figure 5.24 was generated from the data listed in Table 5.12. For this comparison, the image samples were all selected at the data points where CR is approximately 14:1. Figure 5.26 is the image compressed by the 2-way hybrid with CR=13.949 and PSNR=36.698. Figure 5.27 is the image compressed by JPEG with Quality=43, CR=13.67, and PSNR=34.036. Figure 5.28 is the image compressed by fractal with MSE=10, CR=13.755, and PSNR=34.328. The 2-way hybrid eliminates the worst of the problems seen in fractal or JPEG alone [31]. Subjectively, the 2-way hybrid in Figure 5.26 is better than the other methods.

Table 5.12 Compression algorithm results for Kameraman512.pgm image file

Fractal			JPEG			Hybrid	
MSE	CR	PSNR	Quality	CR	PSNR	CR	PSNR
1	2.356	43.100	98	2.358	51.543	2.352	51.544
3.2	4.456	42.080	91	4.448	42.751	4.436	44.338
5.1	7.309	39.406	78	7.291	38.230	7.371	41.544
5.7	8.096	38.604	74	8.036	37.464	8.141	40.782
6.9	9.630	37.199	66	9.632	36.217	9.812	39.314
8.4	11.584	35.700	55	11.587	35.026	11.823	37.910
10	13.755	34.328	43	13.670	34.036	13.949	36.698
13	18.510	32.293	26	18.552	32.033	19.037	34.737
13	19.011	32.109	25	19.008	31.884	19.539	34.588
14	21.174	31.428	21	21.150	31.211	21.801	33.952
15	23.268	30.824	18	23.293	30.645	23.990	33.386
18	29.706	29.457	12	29.717	29.312	30.473	31.917
23	42.942	27.736	6	42.963	27.159	44.216	29.479

Figure 5.25 Original Kameraman512.pgm image.

Figure 5.26 Kameraman512.pgm image using 2-way hybrid $(CR = 13.949, PSNR = 36.698)$.

Figure 5.27 Kameraman512.pgm image using JPEG $(Quality = 43, CR = 13.67, PSNR = 34.036)$.

Figure 5.28 Kameraman512.pgm image using fractal $(MSE = 10, CR = 13.755, PSNR = 34.328)$.

5.13 Results for Lena128.pgm Image

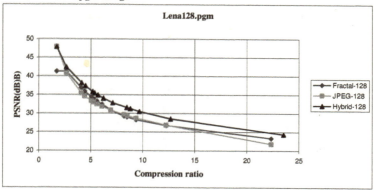

Figure 5.29 Fractal, JPEG, and hybrid compression results for Lena128.pgm.

The graph shown in Figure 5.29 shows an improvement by using the hybrid algorithm over the fractal and JPEG methods. The original image file, shown in part (a) of Figure 5.30, is an 8-bit grayscale image measuring 128x128 pixels. For the fractal algorithm, the compression ratio ranges from 1.733 to 22.342, the PSNR from 41.25 down to 23.289, with an MSE from 1 to 26. The JPEG compression algorithm shows compression ratios varying from 1.734 to 22.342 and PSNR values from 47.869 to 21.789 for quality levels of 98 to 4. Used individually, both fractal and JPEG compression algorithms produce similar results, with the fractal compression algorithm producing the better results. Compressed using the hybrid algorithm compression ratios vary from 1.734 to 23.528 with PSNR values from 47.869 to 24.396. The chart shown in Figure 5.29 was generated from the data listed in Table 5.13. Figure 5.30 shows examples of the Lena image compressed by each method. For this comparison, the image samples were all selected at the data points in Figure 5.29 where CR is approximately 6:1. Part (a) of Figure 5.30 is the original image. Part (b) of Figure 5.30 is the image compressed by the 2-way hybrid with CR=6.224 and PSNR=34.016. Part (c) of Figure 5.30 is the image compressed by JPEG with Quality=49, CR=6.069 and PSNR=31.834. Part (d) of Figure 5.30 is the image compressed by fractal with MSE=10, CR=6.071 and PSNR=32.365. Part (d) exhibits excessive blockiness typical of fractal methods. Clearly part (b) is a much higher quality image having not only a better compression ratio but also a higher PSNR than either part (c) or part (d).

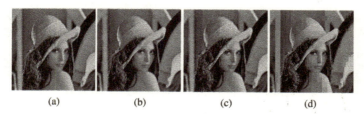

(a)	(b)	(c)	(d)

Figure 5.30 Lena128.pgm image at CR approximately 6:1.

37

Table 5.13 Compression algorithm results for Lena128.pgm image file

Fractal				JPEG				Hybrid	
MSE	CR	PSNR		Quality	CR	PSNR		CR	PSNR
1	1.733	41.250		98	1.734	47.869		1.734	47.869
3.5	2.661	41.234		92	2.667	40.738		2.658	42.369
6.3	4.092	36.902		78	4.094	35.437		4.128	38.155
6.9	4.427	35.888		74	4.424	34.635		4.493	37.343
8.3	5.068	34.356		65	5.068	33.358		5.188	35.817
8.8	5.271	33.929		62	5.276	33.021		5.371	35.459
9.5	5.589	33.257		57	5.585	32.517		5.682	34.887
10	6.071	32.365		49	6.069	31.834		6.224	34.016
12	6.919	30.853		38	6.925	30.837		7.127	32.847
14	8.220	29.247		27	8.200	29.653		8.466	31.561
14	8.501	29.024		25	8.510	29.362		8.802	31.233
15	9.376	28.251		21	9.376	28.674		9.681	30.492
19	12.266	26.729		13	12.256	26.855		12.663	28.534
26	22.342	23.289		4	22.342	21.789		23.528	24.396

5.14 Results for Lena256.pgm Image

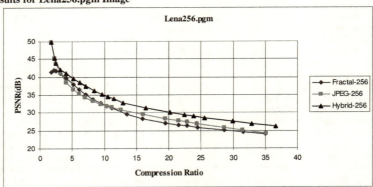

Figure 5.31 Fractal, JPEG, and hybrid compression results for Lena256.pgm.

The graph shown in Figure 5.31 shows an improvement by using the hybrid algorithm over the fractal and JPEG methods. The original image file, shown in part (a) of Figure 5.32, is an 8-bit grayscale image measuring 256x256 pixels. For the fractal algorithm, the compression ratio ranges from 1.756 to 35.0733, the PSNR from 41.361 down to 24.067, with an MSE from 1.2 to 25. The JPEG compression algorithm shows compression ratios varying from 1.754 to 35.092 and PSNR values from 49.882 to 24.165 for quality levels of 98 to 4. Used individually, both fractal and JPEG compression algorithms produce similar results, with the JPEG compression algorithm producing the better results. Compressed using the hybrid algorithm compression ratios vary from 1.751 to 36.682 with PSNR values from 49.882 to 26.096. The chart shown in Figure 5.31 was generated from the data listed in Table 5.14. Figure 5.32 shows examples of the Lena image compressed by each method. For this comparison, the image samples were all selected at the data points in Figure 5.31 where CR is approximately 10:1. Part (a) of Figure 5.32 is the original image. Part (b) of Figure 5.32 is the image compressed by the 2-way hybrid with CR=10.593 and PSNR=34.439. Part (c) of Figure 5.32 is the image compressed by JPEG with Quality=36, CR=10.343 and PSNR=31.815. Part (d) of Figure 5.32 is the image compressed by fractal with MSE=11, CR=10.334 and PSNR=31.753. Part (d) exhibits excessive blockiness typical of fractal

methods. Clearly part (b) is a much higher quality image having not only a better compression ratio but also a higher PSNR than either part (c) or part (d).

Table 5.14 Compression algorithm results for Lena256.pgm image file

Fractal			JPEG			Hybrid	
MSE	CR	PSNR	Quality	CR	PSNR	CR	PSNR
1.2	1.756	41.361	98	1.754	49.882	1.751	49.882
2.3	2.175	41.878	96	2.184	45.159	2.183	45.159
2.8	2.468	41.724	95	2.432	43.723	2.432	43.762
3.6	3.193	40.946	92	3.181	41.122	3.173	42.102
4.3	4.075	39.816	87	4.070	38.591	4.097	41.086
5.3	5.197	37.958	80	5.147	36.569	5.203	39.647
6.2	6.073	36.557	73	6.088	35.284	7.155	37.394
7.2	7.080	35.206	65	7.039	34.285	6.144	38.518
8.5	8.270	33.791	53	8.267	33.159	8.454	36.160
9.6	9.447	32.742	43	9.416	32.402	9.637	35.149
11	10.334	31.753	36	10.343	31.815	10.593	34.439
12	11.135	31.237	32	11.163	31.394	11.440	33.895
14	13.549	29.591	26	12.565	30.754	12.929	32.672
16	15.910	28.247	17	15.992	29.453	16.462	31.333
18	19.532	26.972	12	19.509	28.337	20.251	29.993
19	21.478	26.579	10	21.506	27.701	22.457	29.364
20	22.864	26.302	9	22.864	27.338	23.845	28.972
21	24.459	25.824	8	24.414	26.885	25.556	28.548
23	28.587	25.066	6	28.587	25.813	29.946	27.507
24	31.545	24.510	5	31.485	25.109	32.940	26.883
25	35.073	24.067	4	35.092	24.165	36.682	26.096

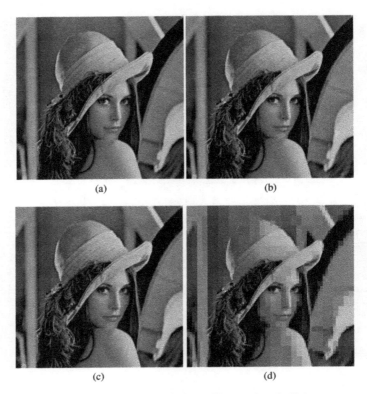

(a)   (b)

(c)   (d)

Figure 5.32 Lena256.pgm image at CR approximately 10:1.

5.15 Results for Lena512.pgm Image

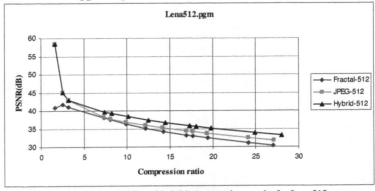

Figure 5.33 Fractal, JPEG, and hybrid compression results for Lena512.pgm.

The graph shown in Figure 5.33 shows an improvement by using the hybrid algorithm over the fractal and JPEG methods. The original image file, shown in Figure 5.34, is an 8-bit grayscale image measuring 512x512 pixels. For the fractal algorithm, the compression ratio ranges from 1.616 to 27.01, the PSNR from 40.806 down to 30.314, with an MSE from 0.2 to 13. The JPEG compression algorithm shows compression ratios varying from 1.616 to 27.13 and PSNR values from 58.478 to 31.678 for quality levels of 100 to 14. Used individually, both fractal and JPEG compression algorithms produce similar results, with the JPEG compression algorithm producing the better results. Compressed using the hybrid algorithm compression ratios vary from 1.611 to 28.056 with PSNR values from 58.478 to 33.166. The chart shown in Figure 5.33 was generated from the data listed in Table 5.15. For this comparison, the image samples were all selected at the data points where CR is approximately 12:1. Figure 5.35 is the image compressed by the 2-way hybrid with CR=12.462 and PSNR=37.431. Figure 5.36 is the image compressed by JPEG with Quality=52, CR=12.205, and PSNR=35.931. Figure 5.37 is the image compressed by fractal with MSE=7, CR=12.21, and PSNR=35.072. The 2-way hybrid eliminates the worst of the problems seen in fractal or JPEG alone [31]. Subjectively, the 2-way hybrid in Figure 5.35 is better than the other methods.

Table 5.15 Compression algorithm results for Lena512.pgm image file

Fractal			JPEG			Hybrid	
MSE	CR	PSNR	Quality	CR	PSNR	CR	PSNR
0.2	1.616	40.806	100	1.616	58.478	1.611	58.478
2.7	2.540	41.769	96	2.548	45.147	2.544	45.147
3.3	3.216	41.100	94	3.213	42.932	3.212	42.992
4.9	7.301	38.029	78	7.299	38.250	7.419	39.705
5.2	8.085	37.517	74	8.090	37.768	8.228	39.306
6	9.923	36.346	65	9.923	36.859	10.097	38.386
7	12.210	35.072	52	12.205	35.931	12.462	37.431
7.8	14.219	34.198	41	14.200	35.239	14.466	36.720
8.9	16.906	33.210	31	16.932	34.375	17.304	35.792
9.2	17.632	32.967	29	17.682	34.163	18.075	35.584
9.9	19.380	32.402	25	19.308	33.698	19.830	35.082
12	24.129	31.133	17	24.113	32.403	24.868	33.851
13	27.010	30.314	14	27.130	31.678	28.056	33.166

Figure 5.34 Original Lena512.pgm image.

Figure 5.35 Lena512.pgm image using 2-way hybrid $(CR = 12.462, PSNR = 37.431)$.

Figure 5.36 Lena512.pgm image using JPEG $(Quality = 52, CR = 12.205, PSNR = 35.931)$.

Figure 5.37 Lena512.pgm image using fractal $(MSE = 7, CR = 12.21, PSNR = 35.072)$.

5.16 Results for Mosaic128.pgm Image

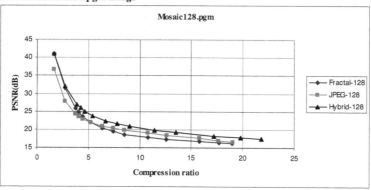

Figure 5.38 Fractal, JPEG, and hybrid compression results for Mosaic128.pgm.

The graph shown in Figure 5.38 shows an improvement by using the hybrid algorithm over the fractal and JPEG methods. The original image file, shown in part (a) of Figure 5.39, is an 8-bit grayscale image measuring 128x128 pixels. For the fractal algorithm, the compression ratio ranges from 1.669 to

18.98, the PSNR from 41.04 down to 16.245, with an MSE from 1.9 to 51.3. The JPEG compression algorithm shows compression ratios varying from 1.668 to 19.046 and PSNR values from 36.595 to 16.626 for quality levels of 90 to 2. Used individually, both fractal and JPEG compression algorithms produce similar results, with the JPEG compression algorithm producing the better results. Compressed using the hybrid algorithm compression ratios vary from 1.661 to 21.807 with PSNR values from 41.05 to 17.419. The chart shown in Figure 5.38 was generated from the data listed in Table 5.16. Figure 5.39 shows examples of the Mosaic image compressed by each method. For this comparison, the image samples were all selected at the data points in Figure 5.38 where CR is approximately 5:1. Part (a) of Figure 5.39 is the original image. Part (b) of Figure 5.39 is the image compressed by the 2-way hybrid with CR=4.64 and PSNR=25.006. Part (c) of Figure 5.39 is the image compressed by JPEG with Quality=33, CR=4.464 and PSNR=22.838. Part (d) of Figure 5.39 is the image compressed by fractal with MSE=26.1, CR=4.465 and PSNR=23.512. Part (d) exhibits excessive blockiness typical of fractal methods. Clearly part (b) is a much higher quality image having not only a better compression ratio but also a higher PSNR than either part (c) or part (d).

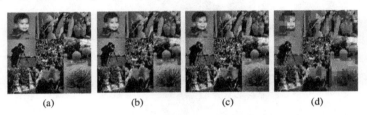

| (a) | (b) | (c) | (d) |

Figure 5.39 Mosaic128.pgm image at CR approximately 5:1.

Table 5.16 Compression algorithm results for Mosaic128.pgm image file

Fractal			JPEG			Hybrid	
MSE	CR	PSNR	Quality	CR	PSNR	CR	PSNR
1.9	1.669	41.040	90	1.668	36.595	1.661	41.050
13.0	2.714	31.467	69	2.716	27.848	2.752	32.046
21.6	3.745	25.764	44	3.747	24.195	3.866	26.895
23.7	4.069	24.717	38	4.071	23.465	4.248	25.979
26.1	4.465	23.512	33	4.464	22.838	4.640	25.006
29.4	5.191	21.980	26	5.194	21.888	5.416	23.667
33.9	6.334	20.283	19	6.334	20.868	6.666	22.279
37.3	7.404	19.463	15	7.414	20.275	7.802	21.505
39.7	8.510	18.609	12	8.506	19.791	8.996	20.862
44.3	10.782	17.787	8	10.789	19.014	11.444	19.795
46.7	12.595	17.288	6	12.605	18.443	13.519	19.201
49.5	15.799	16.730	4	15.783	17.577	17.172	18.222
50.7	17.729	16.500	3	17.614	17.033	19.806	17.763
51.3	18.980	16.245	2	19.046	16.626	21.807	17.419

5.17 Results for Mosaic256.pgm Image

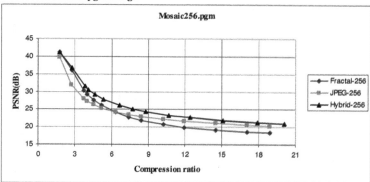

Figure 5.40 Fractal, JPEG, and hybrid compression results for Mosaic256.pgm.

The graph shown in Figure 5.40 shows an improvement by using the hybrid algorithm over the fractal and JPEG methods. The original image file, shown in part (a) of Figure 5.41, is an 8-bit grayscale image measuring 256x256 pixels. For the fractal algorithm, the compression ratio ranges from 1.746 to 18.940, the PSNR from 41.183 down to 18.326, with an MSE from 2.0 to 42.1. The JPEG compression algorithm shows compression ratios varying from 1.745 to 18.902 and PSNR values from 39.817 to 20.207 for quality levels of 93 to 6. Used individually, both fractal and JPEG compression algorithms produce similar results, with the JPEG compression algorithm producing the better results. Compressed using the hybrid algorithm, compression ratios vary from 1.734 to 20.077 with PSNR values from 41.278 to 20.948. The chart shown in Figure 5.40 was generated from the data listed in Table 5.17. Figure 5.41 shows examples of the Mosaic image compressed by each method. For this comparison, the image samples were all selected at the data points in Figure 5.40 where CR is approximately 4:1. Part (a) of Figure 5.41 is the original image. Part (b) of Figure 5.41 is the image compressed by the 2-way hybrid with CR=3.827 and PSNR=31.462. Part (c) of Figure 5.41 is the image compressed by JPEG with Quality=64, CR=3.739 and PSNR=27.929. Part (d) of Figure 5.41 is the image compressed by fractal with MSE=13.6, CR=3.739 and PSNR=30.238. Subjectively, the 2-way hybrid of part (b) is better than the other methods.

Table 5.17 Compression algorithm results for Mosaic256.pgm image file

Fractal			JPEG			Hybrid	
MSE	CR	PSNR	Quality	CR	PSNR	CR	PSNR
2.0	1.746	41.183	93	1.745	39.817	1.734	41.278
7.6	2.747	35.970	80	2.743	31.860	2.767	36.733
13.6	3.739	30.238	64	3.739	27.929	3.827	31.462
15.0	4.013	29.227	59	4.010	27.186	4.127	30.530
17.4	4.519	27.657	51	4.515	26.253	4.650	29.172
20.3	5.153	26.188	41	5.153	25.296	5.338	27.828
24.7	6.356	24.100	30	6.362	24.092	6.645	26.117
27.6	7.398	22.789	24	7.394	23.383	7.732	25.108
29.9	8.420	21.847	20	8.430	22.864	8.783	24.340
33.2	10.230	20.730	15	10.234	22.175	10.673	23.332
35.7	11.916	19.915	12	11.921	21.661	12.417	22.635
38.6	14.499	19.061	9	14.490	21.078	15.093	21.878
40.8	17.044	18.527	7	17.044	20.542	17.930	21.303
42.1	18.940	18.326	6	18.902	20.207	20.077	20.948

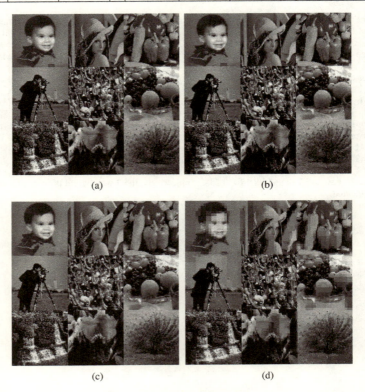

(a)   (b)

(c)   (d)

Figure 5.41 Mosaic256.pgm image at CR approximately 4:1.

48

5.18 Results for Mosaic512.pgm Image

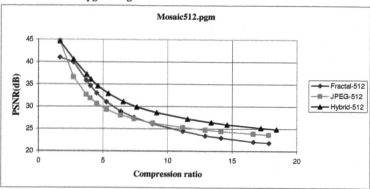

Figure 5.42 Fractal, JPEG, and hybrid compression results for Mosaic512.pgm.

The graph shown in Figure 5.42 shows an improvement by using the hybrid algorithm over the fractal and JPEG methods. The original image file, shown in Figure 5.43, is an 8-bit grayscale image measuring 512x512 pixels. For the fractal algorithm, the compression ratio ranges from 1.660 to 17.845, the PSNR from 40.986 down to 21.86, with an MSE from 0.5 to 31.2. The JPEG compression algorithm shows compression ratios varying from 1.666 to 17.825 and PSNR values from 44.652 to 23.696 for quality levels of 96 to 10. Used individually, both fractal and JPEG compression algorithms produce similar results, with the JPEG compression algorithm producing the better results. Compressed using the hybrid algorithm, compression ratios vary from 1.665 to 18.427 with PSNR values from 44.656 to 24.907. The chart shown in Figure 5.42 was generated from the data listed in Table 5.18. For this comparison, the image samples were all selected at the data points where CR is approximately 4:1. Figure 5.44 is the image compressed by the 2-way hybrid with CR=4.065 and PSNR=36.086. Figure 5.45 is the image compressed by JPEG with Quality=75, CR=4.016, and PSNR=31.82. Figure 5.46 is the image compressed by fractal with MSE=8.4, CR=4.024, and PSNR=34.63. The 2-way hybrid eliminates the worst of the problems seen in fractal or JPEG alone [31]. Subjectively, the 2-way hybrid in Figure 5.44 is better than the other methods.

Table 5.18 Compression algorithm results for Mosaic512.pgm image file

Fractal			JPEG			Hybrid	
MSE	CR	PSNR	Quality	CR	PSNR	CR	PSNR
0.5	1.660	40.986	96	1.666	44.652	1.665	44.656
4.4	2.729	39.955	88	2.719	36.611	2.726	40.572
7.3	3.702	35.863	78	3.701	32.633	3.735	37.203
8.4	4.024	34.630	75	4.016	31.820	4.065	36.086
10.0	4.507	32.949	69	4.513	30.629	4.593	34.615
12.3	5.266	30.990	60	5.270	29.322	5.401	32.874
15.3	6.375	28.915	46	6.376	28.025	6.567	31.049
17.6	7.405	27.527	37	7.403	27.188	7.619	29.860
20.5	8.856	26.117	28	8.854	26.292	9.151	28.608
24.2	11.190	24.460	20	11.184	25.333	11.546	27.206
26.4	12.930	23.449	16	12.931	24.803	13.375	26.428
27.8	14.140	22.992	14	14.142	24.484	14.611	25.979
30.2	16.643	22.117	11	16.638	23.915	17.233	25.202
31.2	17.845	21.860	10	17.825	23.696	18.427	24.907

Figure5.43 Original Mosaic512.pgm image.

Figure 5.44 Mosaic512.pgm image using 2-way hybrid $(CR = 4.065, PSNR = 36.086)$.

Figure 5.45 Mosaic512.pgm image using JPEG $(Quality = 75, CR = 4.016, PSNR = 31.82)$.

Figure 5.46 Mosaic512.pgm image using fractal $(MSE = 8.4, CR = 4.024, PSNR = 34.63)$.

5.19 Results for Mosaic768.pgm Image

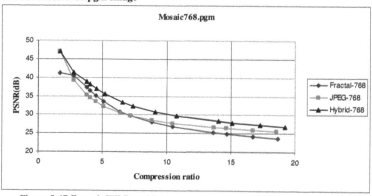

Figure 5.47 Fractal, JPEG, and hybrid compression results for Mosaic768.pgm.

The graph shown in Figure 5.47 shows an improvement by using the hybrid algorithm over the fractal and JPEG methods. The original image file, shown in Figure 5.48, is an 8-bit grayscale image measuring 768x768 pixels. For the fractal algorithm, the compression ratio ranges from 1.689 to 18.652, the PSNR from 41.109 down to 23.563, with an MSE from 0.7 to 26.3. The JPEG compression algorithm shows compression ratios varying from 1.695 to 18.6 and PSNR values from 47.006 to 25.486 for quality levels of 97 to 12. Used individually, both fractal and JPEG compression algorithms produce similar results, with the JPEG compression algorithm producing slightly better results in the end. Compressed using the hybrid algorithm, compression ratios vary from 1.694 to 19.250 with PSNR values from 47.006 to 26.715. The chart shown in Figure 5.47 was generated from the data listed in Table 5.19. For this comparison, the image samples were all selected at the data points where CR is approximately 4:1. Figure 5.49 is the image compressed by the 2-way hybrid with CR=4.04 and PSNR=38.231. Figure 5.50 is the image compressed by JPEG with Quality=81, CR=4.003, and PSNR=34.542. Figure 5.51 is the image compressed by fractal with MSE=6.4, CR=4.009, and PSNR=36.531. The 2-way hybrid eliminates the worst of the problems seen in fractal or JPEG alone [31]. Subjectively, the 2-way hybrid in Figure 5.44 is better than the other methods.

Table 5.19 Compression algorithm results for Mosaic768.pgm image file

Fractal				JPEG				Hybrid	
MSE	CR	PSNR		Quality	CR	PSNR		CR	PSNR
0.700	1.689	41.109		97	1.695	47.006		1.694	47.006
3.900	2.736	40.399		91	2.733	39.209		2.736	41.274
5.800	3.758	37.349		83	3.753	35.196		3.781	38.918
6.400	4.009	36.531		81	4.003	34.542		4.040	38.231
7.600	4.500	35.036		77	4.460	33.448		4.514	36.942
9.100	5.101	33.450		71	5.107	32.202		5.200	35.531
12.000	6.383	30.832		58	6.390	30.417		6.562	33.281
13.500	7.169	29.738		50	7.172	29.669		7.380	32.303
16.300	8.852	27.981		36	8.848	28.474		9.117	30.774
18.500	10.497	26.810		28	10.487	27.679		10.809	29.711
22.100	13.661	25.261		19	13.675	26.634		14.064	28.262
23.100	14.692	24.862		17	14.676	26.359		15.091	27.887
24.900	16.747	24.123		14	16.754	25.870		17.288	27.230
26.300	18.652	23.563		12	18.600	25.486		19.250	26.715

Figure 5.48 Original Mosaic768.pgm image.

Figure 5.49 Mosaic768.pgm image using 2-way hybrid $(CR = 4.040, PSNR = 38.231)$.

Figure 5.50 Mosaic768.pgm image using JPEG $(Quality = 81, CR = 4.003, PSNR = 34.542)$.

Figure 5.51 Mosaic768.pgm image using fractal $(MSE = 6.4, CR = 4.009, PSNR = 36.531)$.

5.20 Results for Orchid128.pgm Image

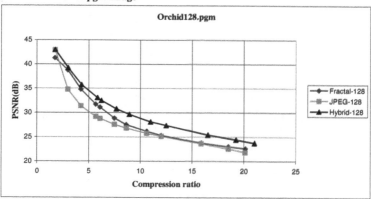

Figure 5.52 Fractal, JPEG, and hybrid compression results for Orchid128.pgm.

     The graph shown in Figure 5.52 shows an improvement by using the hybrid algorithm over the fractal and JPEG methods. The original image file, shown in part (a) of Figure 5.53, is an 8-bit grayscale image measuring 128x128 pixels. For the fractal algorithm, the compression ratio ranges from 1.76 to 20.146, the PSNR from 41.275 down to 22.621, with an MSE from 0.8 to 29.3. The JPEG compression algorithm shows compression ratios varying from 1.761 to 20.121 and PSNR values from 42.894 to 21.838 for quality levels of 95 to 4. Used individually, both fractal and JPEG compression algorithms produce similar results, with the fractal compression algorithm producing slightly better results. Compressed using the hybrid algorithm, compression ratios vary from 1.761 to 21.024 with PSNR values from 42.943 to 23.755. The chart shown in Figure 5.52 was generated from the data listed in Table 5.20. Figure 5.53 shows examples of the Orchid image compressed by each method. For this comparison, the image samples were all selected at the data points in Figure 5.52 where CR is approximately 4:1. Part (a) of Figure 5.53 is the original image. Part (b) of Figure 5.53 is the image compressed by the 2-way hybrid with CR=4.312 and PSNR=35.729. Part (c) of Figure 5.53 is the image compressed by JPEG with Quality=69, CR=4.261 and PSNR=31.339. Part (d) of Figure 5.53 is the image compressed by fractal with MSE=7.6, CR=4.262 and PSNR=34.771. Part (d) exhibits excessive blockiness typical of fractal methods. Clearly part (b) is a much higher quality image having not only a better compression ratio but also a higher PSNR than either part (c) or part (d).

    (a)   (b)   (c)   (d)

Figure 5.53 Orchid128.pgm image at CR approximately 4:1.

Table 5.20 Compression algorithm results for Orchid128.pgm image file

Fractal			JPEG			Hybrid	
MSE	CR	PSNR	Quality	CR	PSNR	CR	PSNR
0.8	1.76	41.275	95	1.761	42.894	1.761	42.943
4.9	2.985	38.786	84	2.987	34.751	3.016	39.23
7.6	4.262	34.771	69	4.261	31.339	4.312	35.729
11	5.676	31.7	47	5.676	29.172	5.846	33.025
12	6.094	31.058	42	6.096	28.762	6.228	32.45
14	7.492	28.835	29	7.492	27.596	7.677	30.749
17	8.649	27.518	22	8.649	26.812	8.942	29.65
20	10.614	26.168	15	10.614	25.764	10.991	28.119
21	12.014	25.316	12	12.014	25.136	12.49	27.377
26	15.875	23.757	7	15.875	23.649	16.515	25.436
28	18.488	23.022	5	18.488	22.576	19.225	24.448
29	20.146	22.621	4	20.121	21.838	21.024	23.755

5.21 Results for Orchid256.pgm Image

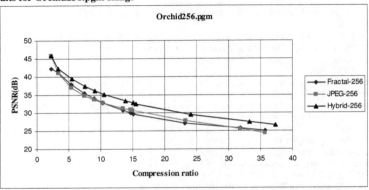

Figure 5.54 Fractal, JPEG, and hybrid compression results for Orchid256.pgm.

The graph shown in Figure 5.54 shows an improvement by using the hybrid algorithm over the fractal and JPEG methods. The original image file, shown in part (a) of Figure 5.55, is an 8-bit grayscale image measuring 256x256 pixels. For the fractal algorithm, the compression ratio ranges from 2.244 to 35.703, the PSNR from 42.265 down to 25.097, with an MSE from 2.1 to 22.6. The JPEG compression algorithm shows compression ratios varying from 2.243 to 35.723 and PSNR values from 45.818 to 24.408 for quality levels of 96 to 4. Used individually, both fractal and JPEG compression algorithms produce similar results, with the JPEG compression algorithm producing slightly better results. Compressed using the hybrid algorithm, compression ratios vary from 2.239 to 37.308 with PSNR values from 45.825 to 26.592. The chart shown in Figure 5.54 was generated from the data listed in Table 5.21. Figure 5.55 shows examples of the Orchid image compressed by each method. For this comparison, the image samples were all selected at the data points in Figure 5.54 where CR is approximately 9:1. Part (a) of Figure 5.55 is the original image. Part (b) of Figure 5.55 is the image compressed by the 2-way hybrid with CR=9.122 and PSNR=36.150. Part (c) of Figure 5.55 is the image compressed by JPEG with Quality=47, CR=8.962 and PSNR=33.746. Part (d) of Figure 5.55 is the image compressed by fractal with MSE=8.2, CR=8.959 and PSNR=33.976. Subjectively, the 2-way hybrid of part (b) is better than the other methods.

Table 5.21 Compression algorithm results for Orchid256.pgm image file

Fractal			JPEG			Hybrid	
MSE	CR	PSNR	Quality	CR	PSNR	CR	PSNR
2.1	2.244	42.265	96	2.243	45.818	2.239	45.825
3.7	3.359	41.093	91	3.352	41.044	3.364	42.229
5.3	5.395	37.800	78	5.390	36.926	5.468	39.412
6.9	7.464	35.440	61	7.454	34.799	7.602	37.383
8.2	8.959	33.976	47	8.962	33.746	9.122	36.150
9.3	10.276	32.891	36	10.273	32.900	10.530	35.160
12	13.513	30.716	23	13.516	31.321	13.841	33.310
13	14.652	30.001	20	14.655	30.849	15.052	32.766
13	15.066	29.776	19	15.048	30.679	15.460	32.549
17	23.163	27.120	9	23.220	27.841	24.073	29.602
21	31.867	25.735	5	31.836	25.488	33.275	27.393
23	35.703	25.097	4	35.723	24.408	37.308	26.592

(a)   (b)

(c)   (d)

Figure 5.55 Orchid256.pgm image at CR approximately 9:1.

61

5.22 Results for Orchid512.pgm Image

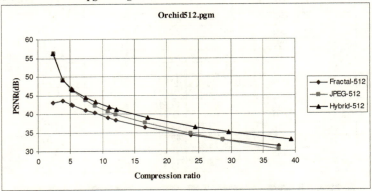

Figure 5.56 Fractal, JPEG, and hybrid compression results for Orchid512.pgm.

The graph shown in Figure 5.56 shows an improvement by using the hybrid algorithm over the fractal and JPEG methods. The original image file, shown in Figure 5.57, is an 8-bit grayscale image measuring 512x512 pixels. For the fractal algorithm, the compression ratio ranges from 2.344 to 37.457, the PSNR from 43.003 down to 31.280, with an MSE from 1.2 to 11.2. The JPEG compression algorithm shows compression ratios varying from 2.399 to 37.489 and PSNR values from 56.316 to 30.539 for quality levels of 99 to 8. Used individually, both fractal and JPEG compression algorithms produce similar results, with the JPEG compression algorithm producing slightly better results. Compressed using the hybrid algorithm, compression ratios vary from 2.375 to 39.428 with PSNR values from 56.316 to 33.051. The chart shown in Figure 5.56 was generated from the data listed in Table 5.22. For this comparison, the image samples were all selected at the data points where CR is approximately 11:1. Figure 5.58 is the image compressed by the 2-way hybrid with CR=11.026 and PSNR=41.794. Figure 5.59 is the image compressed by JPEG with Quality=63, CR=10.877, and PSNR=40.673. Figure 5.60 is the image compressed by fractal with MSE=4.8, CR=10.857, and PSNR=39.053. The 2-way hybrid eliminates the worst of the problems seen in fractal or JPEG alone [31]. Subjectively, the 2-way hybrid in Figure 5.58 is better than the other methods.

Table 5.22 Compression algorithm results for Orchid512.pgm image file

Fractal			JPEG			Hybrid	
MSE	CR	PSNR	Quality	CR	PSNR	CR	PSNR
1.2	1.832	41.221	96	1.838	44.698	1.838	44.698
2.7	2.018	40.887	95	2.019	43.006	2.018	43.018
3.4	3.123	36.253	88	3.123	37.403	3.135	38.051
3.5	5.221	31.453	71	5.212	33.214	5.297	33.716
4	6.380	30.260	60	6.372	32.051	6.497	32.552
4.3	8.928	28.614	37	8.915	30.400	9.137	30.883
4.8	10.721	27.905	28	10.731	29.572	11.055	30.055
5.1	13.932	27.016	20	13.880	28.513	14.460	29.023
6.3	17.524	26.377	15	17.630	27.653	18.509	28.182
8.1	19.939	26.029	13	19.867	27.244	20.953	27.784
9.3	28.071	25.218	9	27.901	26.131	29.964	26.730
11.2	31.161	25.015	8	31.113	25.793	33.632	26.408

Figure 5.57 Original Orchid512.pgm image.

Figure 5.58 Orchid512.pgm image using 2-way hybrid $(CR = 11.026, PSNR = 41.794)$.

Figure 5.59 Orchid512.pgm image using JPEG $(Quality = 63, CR = 10.877, PSNR = 40.673)$.

Figure 5.60 Orchid512.pgm image using fractal $(MSE = 4.8, CR = 10.857, PSNR = 39.053)$.

5.23 Results for Peacock128.pgm Image

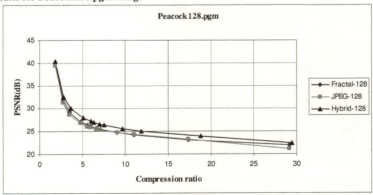

Figure 5.61 Fractal, JPEG, and hybrid compression results for Peacock128.pgm.

The graph shown in Figure 5.61 shows an improvement by using the hybrid algorithm over the fractal and JPEG methods. The original image file, shown in part (a) of Figure 5.62, is an 8-bit grayscale image measuring 128x128 pixels. For the fractal algorithm, the compression ratio ranges from 1.793 to 29.128, the PSNR from 40.120 down to 21.994, with an MSE from 4.3 to 27.1. The JPEG compression algorithm shows compression ratios varying from 1.794 to 29.128 and PSNR values from 39.469 to 21.055 for quality levels of 93 to 2. Used individually, both fractal and JPEG compression algorithms produce similar results, with the fractal compression algorithm producing slightly better results. Compressed using the hybrid algorithm, compression ratios vary from 1.795 to 29.495 with PSNR values from 40.390 to 22.433. The chart shown in Figure 5.61 was generated from the data listed in Table 5.23. Figure 5.62 shows examples of the Peacock image compressed by each method. For this comparison, the image samples were all selected at the data points in Figure 5.61 where CR is approximately 6:1. Part (a) of Figure 5.62 is the original image. Part (b) of Figure 5.62 is the image compressed by the 2-way hybrid with CR=6.341 and PSNR=27.039. Part (c) of Figure 5.62 is the image compressed by JPEG with Quality=41, CR=5.91 and PSNR=25.874. Part (d) of Figure 5.62 is the image compressed by fractal with MSE=17.2, CR=5.91 and PSNR=26.133. Part (d) exhibits excessive blockiness typical of fractal methods. Clearly part (b) is a much higher quality image having not only a better compression ratio but also a higher PSNR than either part (c) or part (d).

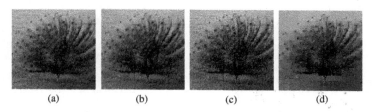

| (a) | (b) | (c) | (d) |

Figure 5.62 Peacock128.pgm image at CR approximately 6:1.

Table 5.23 Compression algorithm results for Peacock128.pgm image file

Fractal			JPEG			Hybrid	
MSE	CR	PSNR	Quality	CR	PSNR	CR	PSNR
4.3	1.793	40.120	93	1.794	39.469	1.795	40.390
10.1	2.756	32.003	81	2.756	31.332	2.808	32.601
12.9	3.524	29.267	71	3.531	28.712	3.690	30.025
15.6	4.762	27.141	55	4.769	26.791	5.072	27.994
16.8	5.582	26.383	44	5.582	26.070	6.003	27.264
17.2	5.910	26.133	41	5.910	25.874	6.341	27.039
18	6.607	25.736	35	6.607	25.491	7.020	26.616
18.6	7.118	25.460	32	7.108	25.303	7.529	26.384
20.3	9.065	24.704	23	9.060	24.761	9.635	25.648
21.7	11.021	24.174	17	11.021	24.263	11.883	25.078
24.5	17.390	23.121	9	17.390	23.193	18.806	23.927
27.1	29.128	21.994	2	29.128	21.055	29.495	22.433

5.24 Results for Peacock256.pgm Image

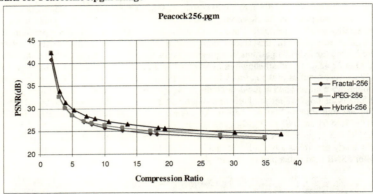

Figure 5.63 Fractal, JPEG, and hybrid compression results for Peacock256.pgm.

The graph shown in Figure 5.63 shows an improvement by using the hybrid algorithm over the fractal and JPEG methods. The original image file, shown in part (a) of Figure 5.64, is an 8-bit grayscale image measuring 256x256 pixels. For the fractal algorithm, the compression ratio ranges from 1.714 to 34.886, the PSNR from 40.860 down to 23.178, with an MSE from 2.8 to 24.9. The JPEG compression algorithm shows compression ratios varying from 1.714 to 34.923 and PSNR values from 42.320 to 23.459 for quality levels of 95 to 6. Used individually, both fractal and JPEG compression algorithms produce similar results, with the JPEG compression algorithm producing slightly better results. Compressed using the hybrid algorithm, compression ratios vary from 1.713 to 37.501 with PSNR values from 42.345 to 24.133. The chart shown in Figure 5.63 was generated from the data listed in Table 5.24. Figure 5.64 shows examples of the Peacock image compressed by each method. For this comparison, the image samples were all selected at the data points in Figure 5.63 where CR is approximately 5:1. Part (a) of Figure 5.64 is the original image. Part (b) of Figure 5.64 is the image compressed by the 2-way hybrid with CR=5.296 and PSNR=29.731. Part (c) of Figure 5.64 is the image compressed by JPEG with Quality=63, CR=5.011 and PSNR=28.522. Part (d) of Figure 5.64 is the image compressed by fractal with MSE=13.4, CR=5.009 and PSNR=28.547. Subjectively, the 2-way hybrid of part (b) is better than the other methods.

Table 5.24 Compression algorithm results for Peacock256.pgm image file

Fractal			JPEG			Hybrid	
MSE	CR	PSNR	Quality	CR	PSNR	CR	PSNR
2.8	1.714	40.860	95	1.714	42.320	1.713	42.345
9	2.994	32.817	83	2.990	32.646	3.051	33.851
11.3	3.830	30.331	75	3.835	30.208	4.002	31.431
13.4	5.009	28.547	63	5.011	28.522	5.296	29.731
15.6	6.830	27.099	45	6.822	27.288	7.256	28.384
16.6	8.044	26.459	36	8.043	26.800	8.512	27.781
18.1	10.091	25.750	27	10.094	26.244	10.652	27.119
19.5	12.818	25.102	20	12.808	25.698	13.555	26.485
21.1	17.133	24.413	14	17.115	25.067	18.331	25.746
21.4	18.148	24.315	13	18.118	24.946	19.423	25.609
23.8	28.013	23.571	8	28.037	24.038	30.208	24.647
24.9	34.886	23.178	6	34.923	23.459	37.501	24.133

(a)   (b)

(c)   (d)

Figure 5.64 Peacock256.pgm image at CR approximately 5:1.

5.25 Results for Peacock512.pgm Image

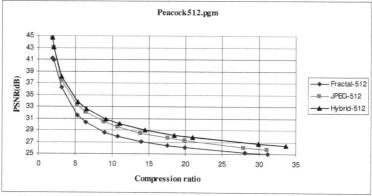

Figure 5.65 Fractal, JPEG, and hybrid compression results for Peacock512.pgm.

The graph shown in Figure 5.65 shows an improvement by using the hybrid algorithm over the fractal and JPEG methods. The original image file, shown in Figure 5.66, is an 8-bit grayscale image measuring 512x512 pixels. For the fractal algorithm, the compression ratio ranges from 1.832 to 31.161, the PSNR from 41.221 down to 25.015, with an MSE from 2.4 to 19.7. The JPEG compression algorithm shows compression ratios varying from 1.838 to 31.113 and PSNR values from 44.698 to 25.793 for quality levels of 96 to 8. Used individually, both fractal and JPEG compression algorithms produce similar results, with the JPEG compression algorithm producing slightly better results. Compressed using the hybrid algorithm, compression ratios vary from 1.838 to 33.632 with PSNR values from 44.698 to 26.408. The chart shown in Figure 5.65 was generated from the data listed in Table 5.25. For this comparison, the image samples were all selected at the data points where CR is approximately 9:1. Figure 5.67 is the image compressed by the 2-way hybrid with CR=9.137 and PSNR=30.883. Figure 5.68 is the image compressed by JPEG with Quality=37, CR=8.915, and PSNR=30.4. Figure 5.69 is the image compressed by fractal with MSE=12.8, CR=8.928, and PSNR=28.614. The 2-way hybrid eliminates the worst of the problems seen in fractal or JPEG alone [31]. Subjectively, the 2-way hybrid in Figure 5.67 is better than the other methods.

Table 5.25 Compression algorithm results for Peacock512.pgm image file

Fractal			JPEG			Hybrid	
MSE	CR	PSNR	Quality	CR	PSNR	CR	PSNR
2.4	1.832	41.221	96	1.838	44.698	1.838	44.698
3.2	2.018	40.887	95	2.019	43.006	2.018	43.018
5.9	3.123	36.253	88	3.123	37.403	3.135	38.051
9.4	5.221	31.453	71	5.212	33.214	5.297	33.716
10.7	6.380	30.260	60	6.372	32.051	6.497	32.552
12.8	8.928	28.614	37	8.915	30.400	9.137	30.883
13.9	10.721	27.905	28	10.731	29.572	11.055	30.055
15.4	13.932	27.016	20	13.880	28.513	14.460	29.023
16.7	17.524	26.377	15	17.630	27.653	18.509	28.182
17.4	19.939	26.029	13	19.867	27.244	20.953	27.784
19.2	28.071	25.218	9	27.901	26.131	29.964	26.730
19.7	31.161	25.015	8	31.113	25.793	33.632	26.408

Figure 5.66 Original Peacock512.pgm image.

Figure 5.67 Peacock512.pgm image using 2-way hybrid$(CR = 9.137, PSNR = 30.883)$.

Figure 5.68 Peacock512.pgm image using JPEG $(Quality = 37, CR = 8.915, PSNR = 30.400)$.

Figure 5.69 Peacock512.pgm image using fractal $(MSE = 12.8, CR = 8.928, PSNR = 28.614)$.

5.26 Results for Peppers128.pgm Image

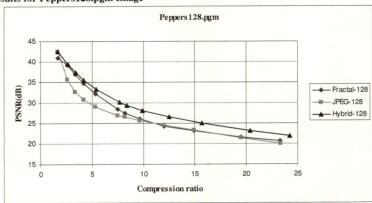

Figure 5.70 Fractal, JPEG, and hybrid compression results for Peppers128.pgm.

The graph shown in Figure 5.70 shows an improvement by using the hybrid algorithm over the fractal and JPEG methods. The original image file, shown in part (a) of Figure 5.71, is an 8-bit grayscale

74

image measuring 128x128 pixels. For the fractal algorithm, the compression ratio ranges from 1.666 to 23.294, the PSNR from 40.995 down to 20.625, with an MSE from 1.5 to 35.1. The JPEG compression algorithm shows compression ratios varying from 1.669 to 23.327 and PSNR values from 42.464 to 19.945 for quality levels of 95 to 2. Used individually, both fractal and JPEG compression algorithms produce similar results, with the fractal compression algorithm producing slightly better results. Compressed using the hybrid algorithm, compression ratios vary from 1.670 to 24.259 with PSNR values from 42.486 to 21.873. The chart shown in Figure 5.70 was generated from the data listed in Table 5.26. Figure 5.71 shows examples of the Peppers image compressed by each method. For this comparison, the image samples were all selected at the data points in Figure 5.70 where CR is approximately 8:1. Part (a) of Figure 5.71 is the original image. Part (b) of Figure 5.71 is the image compressed by the 2-way hybrid with CR=7.695 and PSNR=30.157. Part (c) of Figure 5.71 is the image compressed by JPEG with Quality=27, CR=7.471 and PSNR=26.939. Part (d) of Figure 5.71 is the image compressed by fractal with MSE=16.3, CR=7.474 and PSNR=28.457. Part (d) exhibits excessive blockiness typical of fractal methods. Clearly part (b) is a much higher quality image having not only a better compression ratio but also a higher PSNR than either part (c) or part (d).

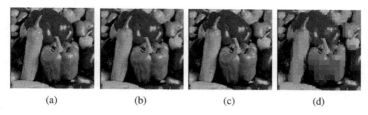

(a)   (b)   (c)   (d)

Figure 5.71 Peppers128.pgm image at CR approximately 8:1.

Table 5.26 Compression algorithm results for Peppers128.pgm image file

Fractal			JPEG			Hybrid	
MSE	CR	PSNR	Quality	CR	PSNR	CR	PSNR
1.5	1.666	40.995	95	1.669	42.464	1.670	42.486
4.5	2.598	39.199	87	2.598	35.639	2.634	39.442
5.9	3.360	36.819	78	3.370	32.685	3.417	37.367
7.5	4.162	34.686	68	4.161	30.853	4.183	35.491
10.4	5.299	32.212	50	5.299	29.037	5.402	33.305
16.3	7.474	28.457	27	7.471	26.939	7.695	30.157
18.1	8.167	27.545	23	8.171	26.505	8.405	29.417
21.3	9.629	26.024	17	9.629	25.694	9.927	28.165
24.9	11.988	24.273	11	11.988	24.568	12.499	26.509
28.1	14.935	23.100	7	14.895	23.255	15.648	24.966
32.2	19.476	21.554	4	19.453	21.390	20.371	23.161
35.1	23.294	20.625	2	23.327	19.945	24.259	21.873

5.27 Results for Peppers256.pgm Image

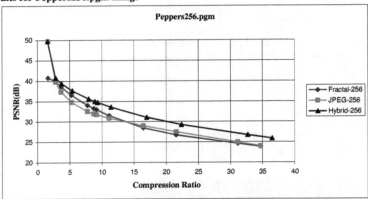

Figure 5.72 Fractal, JPEG, and hybrid compression results for Peppers256.pgm.

The graph shown in Figure 5.72 shows an improvement by using the hybrid algorithm over the fractal and JPEG methods. The original image file, shown in part (a) of Figure 5.73, is an 8-bit grayscale image measuring 256x256 pixels. For the fractal algorithm, the compression ratio ranges from 1.610 to 34.701, the PSNR from 40.798 down to 23.865, with an MSE from 0.3 to 26.1. The JPEG compression algorithm shows compression ratios varying from 1.610 to 34.701 and PSNR values from 49.775 to 23.992 for quality levels of 98 to 4. Used individually, both fractal and JPEG compression algorithms produce similar results, with the fractal compression algorithm producing slightly better results. Compressed using the hybrid algorithm, compression ratios vary from 1.609 to 36.478 with PSNR values from 49.775 to 25.911. The chart shown in Figure 5.72 was generated from the data listed in Table 5.27. Figure 5.73 shows examples of the Peppers image compressed by each method. For this comparison, the image samples were all selected at the data points in Figure 5.72 where CR is approximately 9:1. Part (a) of Figure 5.73 is the original image. Part (b) of Figure 5.73 is the image compressed by the 2-way hybrid with CR=8.922 and PSNR=34.981. Part (c) of Figure 5.73 is the image compressed by JPEG with Quality=45, CR=8.758 and PSNR=31.950. Part (d) of Figure 5.73 is the image compressed by fractal with MSE=8.7, CR=8.758 and PSNR=33.303. Subjectively, the 2-way hybrid of part (b) is better than the other methods.

Table 5.27 Compression algorithm results for Peppers256.pgm image file

Fractal			JPEG			Hybrid	
MSE	CR	PSNR	Quality	CR	PSNR	CR	PSNR
0.3	1.610	40.798	98	1.610	49.775	1.609	49.775
3.9	2.826	39.988	92	2.816	39.792	2.836	40.709
4.6	3.665	38.587	87	3.671	37.294	3.737	39.405
5.8	5.283	36.490	76	5.289	34.786	5.394	37.613
7.8	7.787	34.039	54	7.787	32.529	7.947	35.605
8.7	8.758	33.303	45	8.758	31.950	8.922	34.981
9.1	9.159	32.985	42	9.156	31.737	9.336	34.735
11	11.112	31.515	31	11.110	30.879	11.400	33.592
16.1	16.449	28.470	16	16.441	28.968	16.987	31.067
19.9	21.556	26.727	10	21.556	27.488	22.372	29.254
24.7	31.141	24.628	5	31.111	24.948	32.694	26.767
26.1	34.701	23.865	4	34.701	23.992	36.478	25.911

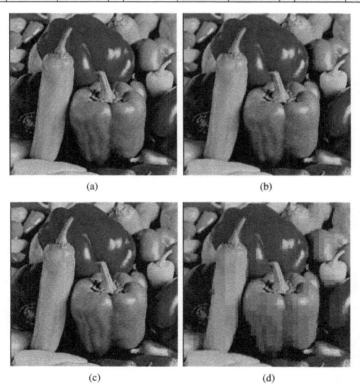

(a)   (b)

(c)   (d)

Figure 5.73 Peppers256.pgm image at CR approximately 9:1.

5.28 Results for Peppers512.pgm Image

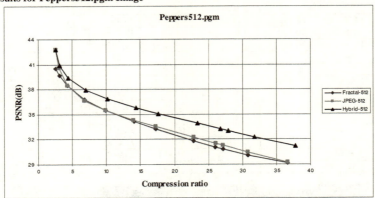

Figure 5.74 Fractal, JPEG, and hybrid compression results for Peppers512.pgm.

     The graph shown in Figure 5.74 shows an improvement by using the hybrid algorithm over the fractal and JPEG methods. The original image file, shown in Figure 5.75, is an 8-bit grayscale image measuring 512x512 pixels. For the fractal algorithm, the compression ratio ranges from 2.547 to 36.630, the PSNR from 40.506 down to 29.173, with an MSE from 3.4 to 15.2. The JPEG compression algorithm shows compression ratios varying from 2.544 to 36.604 and PSNR values from 42.789 to 29.212 for quality levels of 95 to 8. Used individually, both fractal and JPEG compression algorithms produce similar results, with the JPEG compression algorithm producing slightly better results. Compressed using the hybrid algorithm, compression ratios vary from 1.609 to 36.478 with PSNR values from 49.775 to 25.911. The chart shown in Figure 5.74 was generated from the data listed in Table 5.28. For this comparison, the image samples were all selected at the data points where CR is approximately 14:1. Figure 5.76 is the image compressed by the 2-way hybrid with CR=14.378 and PSNR=35.811. Figure 5.77 is the image compressed by JPEG with Quality=41, CR=14.038, and PSNR=34.292. Figure 5.78 is the image compressed by fractal with MSE=7.5, CR=14.061, and PSNR=34.147. The 2-way hybrid eliminates the worst of the problems seen in fractal or JPEG alone [31]. Subjectively, the 2-way hybrid in Figure 5.76 is better than the other methods.

Table 5.28 Compression algorithm results for Peppers512.pgm image file

Fractal			JPEG			Hybrid	
MSE	CR	PSNR	Quality	CR	PSNR	CR	PSNR
3.4	2.547	40.506	95	2.544	42.789	2.542	42.801
3.9	3.118	39.706	93	3.114	40.604	3.133	40.870
4.5	4.285	38.472	89	4.274	38.446	4.416	39.399
5.4	6.842	36.771	79	6.852	36.657	7.037	37.967
6.3	9.884	35.484	64	9.880	35.493	10.135	36.911
7.5	14.061	34.147	41	14.038	34.292	14.378	35.811
8.5	17.302	33.264	30	17.313	33.533	17.684	35.096
10.5	22.860	31.775	19	22.804	32.260	23.395	33.901
11.7	26.041	31.046	15	26.096	31.529	26.762	33.250
12.1	27.144	30.808	14	27.170	31.292	27.883	33.013
13.4	30.788	30.045	11	30.806	30.461	31.750	32.278
15.2	36.630	29.173	8	36.604	29.212	37.748	31.219

Figure 5.75 Original Peppers512.pgm image.

Figure 5.76 Peppers512.pgm image using 2-way hybrid $(CR = 14.378, PSNR = 35.811)$.

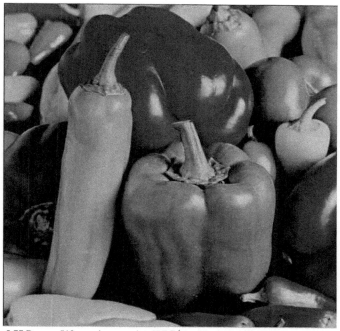

Figure 5.77 Peppers512.pgm image using JPEG $(Quality = 41, CR = 14.038, PSNR = 34.292)$.

Figure 5.78 Peppers512.pgm image using fractal $(MSE = 7.5, CR = 14.061, PSNR = 34.147)$.

5.29 Results for Waterfall128.pgm Image

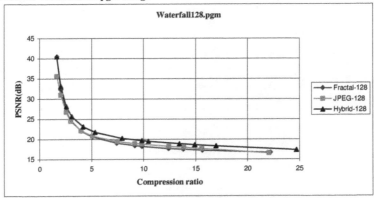

Figure 5.79 Fractal, JPEG, and hybrid compression results for Waterfall128.pgm.

The graph shown in Figure 5.79 shows an improvement by using the hybrid algorithm over the fractal and JPEG methods. The original image file, shown in part (a) of Figure 5.80, is an 8-bit grayscale image measuring 128x128 pixels. For the fractal algorithm, the compression ratio ranges from 1.651 to 22.191, the PSNR from 40.503 down to 16.686, with an MSE from 4.5 to 47.9. The JPEG compression algorithm shows compression ratios varying from 1.651 to 22.042 and PSNR values from 35.572 to 16.517 for quality levels of 89 to 2. Used individually, both fractal and JPEG compression algorithms produce similar results, with the JPEG compression algorithm producing slightly better results. Compressed using the hybrid algorithm, compression ratios vary from 1.649 to 24.697 with PSNR values from 40.503 to 17.326. The chart shown in Figure 5.79 was generated from the data listed in Table 5.29. Figure 5.80 shows examples of the Waterfall image compressed by each method. For this comparison, the image samples were all selected at the data points in Figure 5.79 where CR is approximately 4:1. Part (a) of Figure 5.80 is the original image. Part (b) of Figure 5.80 is the image compressed by the 2-way hybrid with CR=4.182 and PSNR=23.139. Part (c) of Figure 5.80 is the image compressed by JPEG with Quality=37, CR=3.982 and PSNR=22.133. Part (d) of Figure 5.80 is the image compressed by fractal with MSE=27.6, CR=3.978 and PSNR=22.075. Part (d) exhibits excessive blockiness typical of fractal methods. Clearly part (b) is a much higher quality image having not only a better compression ratio but also a higher PSNR than either part (c) or part (d).

| (a) | (b) | (c) | (d) |

Figure 5.80 Waterfall128.pgm image at CR approximately 4:1.

Table 5.29 Compression algorithm results for Waterfall128.pgm image file

Fractal			JPEG			Hybrid	
MSE	CR	PSNR	Quality	CR	PSNR	CR	PSNR
4.5	1.651	40.503	89	1.651	35.572	1.649	40.503
12.1	2.022	32.330	81	2.022	30.952	2.030	33.096
18.2	2.545	27.086	68	2.545	26.725	2.584	28.170
22.2	3.002	24.689	56	3.003	24.509	3.099	25.684
27.6	3.978	22.075	37	3.982	22.133	4.182	23.139
31.9	5.094	20.612	26	5.094	20.782	5.336	21.742
37.3	7.427	19.033	15	7.434	19.445	7.926	20.256
39.9	9.161	18.424	11	9.146	18.922	9.843	19.655
40.6	9.855	18.204	10	9.826	18.767	10.459	19.461
43.2	12.414	17.668	7	12.433	18.193	13.442	18.871
44.2	13.827	17.490	6	13.851	17.948	14.908	18.588
45.3	15.648	17.244	5	15.588	17.666	16.959	18.284
47.9	22.191	16.686	2	22.042	16.517	24.697	17.326

5.30 Results for Waterfall256.pgm Image

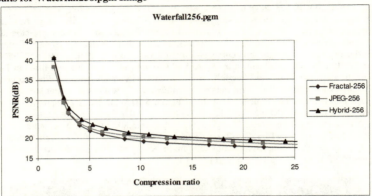

Figure 5.81 Fractal, JPEG, and hybrid compression results for Waterfall256.pgm.

The graph shown in Figure 5.81 shows an improvement by using the hybrid algorithm over the fractal and JPEG methods. The original image file, shown in part (a) of Figure 5.82, is an 8-bit grayscale image measuring 256x256 pixels. For the fractal algorithm, the compression ratio ranges from 1.604 to 32.307, the PSNR from 40.759 down to 17.032, with an MSE from 0.7 to 46.5. The JPEG compression algorithm shows compression ratios varying from 1.604 to 32.307 and PSNR values from 38.378 to 17.658 for quality levels of 92 to 3. Used individually, both fractal and JPEG compression algorithms produce similar results, with the JPEG compression algorithm producing slightly better results. Compressed using the hybrid algorithm, compression ratios vary from 1.602 to 36.621 with PSNR values from 40.759 to 18.150. The chart shown in Figure 5.81 was generated from the data listed in Table 5.30. Figure 5.82 shows examples of the Waterfall image compressed by each method. For this comparison, the image samples were all selected at the data points in Figure 5.81 where CR is approximately 4:1. Part (a) of Figure 5.82 is the original image. Part (b) of Figure 5.82 is the image compressed by the 2-way hybrid with CR=4.267 and PSNR=24.916. Part (c) of Figure 5.82 is the image compressed by JPEG with Quality=45, CR=4.095 and PSNR=23.879. Part (d) of Figure 5.82 is the image compressed by fractal with MSE=23.5, CR=4.096 and PSNR=23.479. Subjectively, the 2-way hybrid of part (b) is better than the other methods.

Table 5.30 Compression algorithm results for Waterfall256.pgm image file

Fractal			JPEG			Hybrid	
MSE	CR	PSNR	Quality	CR	PSNR	CR	PSNR
0.7	1.604	40.759	92	1.604	38.378	1.602	40.759
14.3	2.520	29.225	76	2.519	29.254	2.541	30.548
17.9	3.005	26.473	66	3.011	26.742	3.063	27.869
23.5	4.096	23.479	45	4.095	23.879	4.267	24.916
27.1	5.102	22.047	33	5.107	22.572	5.362	23.526
30	6.323	20.983	25	6.318	21.693	6.621	22.578
33.7	8.482	19.865	17	8.480	20.823	8.853	21.545
35.8	10.334	19.251	13	10.329	20.325	10.803	20.970
38	12.582	18.761	10	12.589	19.850	13.291	20.425
40.8	16.667	18.179	7	16.718	19.220	18.083	19.706
42	19.117	17.882	6	18.995	18.936	20.672	19.381
43.4	22.019	17.646	5	21.923	18.585	24.100	19.015
46.5	32.307	17.032	3	32.307	17.658	36.621	18.150

(a)

(b)

(c)

(d)

Figure 5.82 Waterfall256.pgm image at CR approximately 4:1.

5.31 Results for Waterfall512.pgm Image

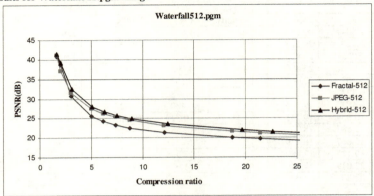

Figure 5.83 Fractal, JPEG, and hybrid compression results for Waterfall512.pgm.

The graph shown in Figure 5.83 shows an improvement by using the hybrid algorithm over the fractal and JPEG methods. The original image file, shown in Figure 5.84, is an 8-bit grayscale image measuring 512x512 pixels. For the fractal algorithm, the compression ratio ranges from 1.642 to 38.182, the PSNR from 40.855 down to 18.438, with an MSE from 1.7 to 39.9. The JPEG compression algorithm shows compression ratios varying from 1.641 to 38.333 and PSNR values from 41.083 to 19.199 for quality levels of 94 to 3. Used individually, both fractal and JPEG compression algorithms produce similar results, with the JPEG compression algorithm producing slightly better results. Compressed using the hybrid algorithm, compression ratios vary from 1.637 to 42.489 with PSNR values from 41.370 to 19.780. The chart shown in Figure 5.83 was generated from the data listed in Table 5.31. For this comparison, the image samples were all selected at the data points where CR is approximately 5:1. Figure 5.85 is the image compressed by the 2-way hybrid with CR=5.095 and PSNR=27.919. Figure 5.86 is the image compressed by JPEG with Quality=46, CR=5.010, and PSNR=27.417. Figure 5.87 is the image compressed by fractal with MSE=17.7, CR=4.998, and PSNR=25.575. The 2-way hybrid eliminates the worst of the problems seen in fractal or JPEG alone [31]. Subjectively, the 2-way hybrid in Figure 5.85 is better than the other methods.

Table 5.31 Compression algorithm results for Waterfall512.pgm image file

Fractal			JPEG			Hybrid	
MSE	CR	PSNR	Quality	CR	PSNR	CR	PSNR
1.7	1.642	40.855	94	1.641	41.083	1.637	41.370
5.1	2.029	38.690	90	2.030	37.052	2.026	39.139
10.7	3.051	30.732	77	3.056	31.513	3.083	32.460
17.7	4.998	25.575	46	5.010	27.417	5.095	27.919
20.5	6.171	24.204	33	6.172	26.203	6.293	26.654
22.8	7.367	23.265	25	7.365	25.314	7.509	25.724
24.9	8.775	22.504	19	8.784	24.510	8.971	24.917
28.6	12.093	21.312	12	12.068	23.173	12.449	23.550
33.3	18.711	20.002	7	18.712	21.586	19.642	21.985
34.7	21.474	19.640	6	21.427	21.139	22.637	21.551
36.2	25.232	19.249	5	25.212	20.621	26.817	21.046
38	30.466	18.856	4	30.441	19.971	33.034	20.445
39.9	38.182	18.438	3	38.333	19.199	42.489	19.780

Figure 5.84 Original Waterfall512.pgm image.

Figure 5.85 Waterfall512.pgm image using 2-way hybrid $(CR = 5.095, PSNR = 27.919)$.

Figure 5.86 Waterfall512.pgm image using JPEG $(Quality = 46, CR = 5.010, PSNR = 27.417)$.

Figure 5.87 Waterfall512.pgm image using fractal $(MSE = 17.7, CR = 4.998, PSNR = 25.575)$.

5.32 Conclusions
      As demonstrated with thirty test images, the 2-way hybrid image compression method exceeds the capabilities of both fractal and JPEG compression methods alone. PSNR improvements in the range of 2-3 dB are common with all of the test images at any given CR. Similarily, CR improvements in the range of 40-50% for any given PSNR are also common. Because of the restructuring of the fractal algorithm from an $O(n^4)$ traditional search-based algorithm to a faster $O(n^2)$ searchless algorithm, the execution speed of the 2-way hybrid algorithm is not a significant limiting factor [23]. The compression of most of the test images takes only a few seconds running in a MATLAB environment on a Windows XP based computer.

CHAPTER 6
JPEG2000

6.1 JPEG2000 Image Compression

Different from other image coding schemes, JPEG2000 supports both lossless and lossy image compression, depending on the wavelet transform and quantization scheme applied. The JPEG2000 image compression standard is based on partitioning the image into smaller subdivisions, called tiles or blocks, which are treated as separate, distinct images, and are compressed independently of each other. This represents the strongest form of spatial partitioning. All image compression operations including wavelet transforms, quantization, and entropy coding, are performed independently of each other on all of the different blocks in the image. Every block in the image has the exact same dimension, with the exception being those blocks residing in the lower left and right sides of the image. Varying block sizes are acceptable up to and including the image as a whole. Subdividing the image into blocks greatly reduces the amount of memory required for execution of the image compression algorithm. In addition, image subdivision is one of the more efficient extraction methods for the numerous regions that comprise an image.

The block diagram for the JPEG2000 encoder is shown in Figure 6.1a. First, a discrete wavelet transform (DWT) is applied to the image source data. Next, the transform coefficients are quantized and then entropy encoded, making up the bitstream output. Shown in Figure 6.1b is the block diagram for the JPEG2000 decoder. In this process the bitstream data is entropy decoded, then dequantized, and finally inverse discrete transformed. The result of this process yields the reconstructed image data.

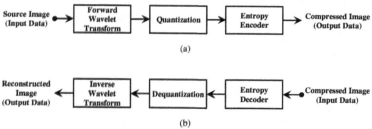

(a)

(b)

Figure 6.1 JPEG2000 block diagram: (a) encoder and (b) decoder.

6.2 Wavelet Transformation Process

Before computing the forward DWT on each block, each image block component is DC level shifted. As in the original JPEG standard, the pixels are level shifted by subtracting 2^{m-1}, where 2^m is the number of gray levels present in the image. If the image has more than one component, like the red, green, and blue planes of a color image, then each component is individually shifted [1]. Using a 1-D wavelet transform, the various block components are broken down into different levels, which contain subbands whose coefficients define both the horizontal and vertical spatial frequency characteristics of the original block component planes. Repeating this process N_L times, with all subsequent iterations being restricted to the previous decomposition's approximation coefficients, will produce an N_L-scale wavelet transform [11]. This process is shown in Figure 6.2. These different levels of decomposition are related to each other by a spatial power of two.

91

Image Blocks   DWT Blocks

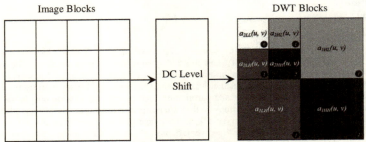

Figure 6.2 JPEG2000 wavelet transformation coefficient notation.

To perform the forward DWT, the standard uses a 1-D subband decomposition of a 1-D set of samples into low-pass samples, representing a downsampled low-resolution version of the original set, and high-pass samples, representing a downsampled residual version of the original set, needed for the perfect reconstruction of the original set from the low-pass set [2]. Generally, any user supplied wavelet filter bank may be used. However, for error-free compression, the wavelet transform used is biorthogonal, with a 5-3 coefficient scaling and wavelet vector and for lossy compression, a 9-7 coefficient scaling and wavelet vector is used [11]. The DWT process can be either reversible or irreversible.

6.3 Quantization

The term quantization refers to the process of reducing the precision of the wavelet transform coefficients. After computing the N_L-scale wavelet transform, the total number of transformation coefficients will equal the number of samples from the original image. Unless a quantization step of 1 is used and all the coefficients are integers, the result of this operation will be lossy. In order to reduce the number of bits needed to represent the transform coefficients $a_b(u,v)$ of subband b, each one is quantized to the value $q_b(u,v)$ (i.e. scalar quantization) using the following equation:

$$q_b(u,v) = \text{sign}\left[a_b(u,v)\right] \cdot \left\lfloor \frac{\left|a_b(u,v)\right|}{\Delta_b} \right\rfloor \qquad (6.1)$$

where Δ_b is the quantization step size of subband b and is determined by Eq. (6.2).

$$\Delta_b = 2^{R_b - \varepsilon_b} \cdot \left(1 + \frac{\mu_b}{2^{11}}\right)$$

R_b = nominal dynamic range of subband b $\qquad (6.2)$
ε_b = # of bits allotted for exponent of subband's coefficient
μ_b = # of bits allotted for mantissa of subband's coefficient

The nominal dynamic range for subband b is equal to the sum of the number of bits used to comprise the original image and the analysis gain bits for subband b [1, 11]. Figure 6.3 shows the pattern that the subband analysis gain bits follow.

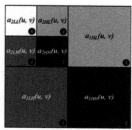

Figure 6.3 JPEG2000 Subband structure.

Only one quantization step per subband is allowed and all quantized transform coefficients will be signed values even if the original component values are unsigned. These coefficients are expressed in a sign-magnitude representation prior to being coded as given by Eq. (6.2).

For error-free image compression, the number of bits allotted for the mantissa of each subband's coefficient, μ_b must be zero. In addition, the number of bits allotted for the exponent of each subband's coefficient, ε_b must equal the nominal dynamic range of subband b (i.e. $R_b = \varepsilon_b$). Substituting these values in Eq. (6.2) will result in a value of one for Δ_b ($\Delta_b = 1$). There is no specific quantization step size defined under the JPEG2000 standard for irreversible compression. The number of bits representing the exponent and mantissa must be provided by the user to the decoder either on a subband basis, called explicit quantization, or for the $N_L LL$ subband only, known as implicit quantization. With implicit quantization, the remaining subbands are quantized using only the extrapolated $N_L LL$ subband parameters. Choosing ε_0 and μ_0 to represent the number of bits reserved for the $N_L LL$ subband, the extrapolated parameters for subband b are then represented by the following set of equations

$$\mu_b = \mu_0$$
$$\varepsilon_b = \varepsilon_0 + nsd_b - nsd_0 \tag{6.3}$$

where nsd_b is the number of subband decomposition levels from the original image tile component to subband b [11]. The final step of the encoding process is to code the quantized coefficients arithmetically on a bit-plane basis. Arithmetic coding is a variable-length coding procedure and like Huffman coding is designed to reduce coding redundancy [1, 11].

6.4 Entropy Coding

The JPEG 2000 entropy coding process of Figure 6.1 is only an approximation. For this part of the algorithm, Huffman encoding followed up with zero run-length coding is used. This is done for simplicity sake only. The Matlab code, im2jpeg2k.m, to implement this process is given in Appendix B and provided courtesy of Gonzalez, et al. [11].

The first step in the process of Huffman coding is to create a series of source reductions by ordering the probabilities of the symbols under consideration and combining the lowest probability symbols into a single symbol that replaces them in the next source reduction [1]. Figure 6.4 shows the process of binary coding. In the leftmost column are the hypothetical sets of source symbols used to illustrate the generation of Huffman codes. The probabilities are in order from the top to the bottom in decreasing value. For the first Huffman source reduction, the last two probabilities, 0.07 and 0.03, are summed together to form a combined symbol having a probability of 0.1. This combined probability value is then placed at the bottom of the first source reduction column to keep the reduced source probabilities ordered from top to bottom in decreasing probability value. Repetition of this process yields the two reduced source probabilities shown in column four.

	Original source		Source reduction		
Symbol	Probability	1	2	3	4
h_6	0.50	0.50	0.50	0.50	0.50
h_3	0.20	0.20	0.20	0.30	0.50
h_4	0.15	0.15	0.20	0.20	
h_5	0.10	0.10	0.10		
h_1	0.07	0.10			
h_2	0.03				

Figure 6.4 Huffman source reductions example.

The next step is to code each of the reduced sources, beginning with the smallest source value and then working back to the original source. Coding these values in a binary code, there are only two possible values for each symbol and those are 0 and 1. As shown in Figure 6.5, symbol values are assigned starting with the values listed in column four. The assignment of the symbol values chosen is strictly arbitrary; reversing the order of the assignment work also work too. The probability value of 0.50 in column four was obtained from the combination of the two reduced source symbols to the left in column three. The 0 used to code this value now gets reassigned to both symbols and new values of 0 and 1 are arbitrarily chosen and added to each of them to aid in distinguishing them from one another.

	Original source			Source reduction							
Symbol	Probability	Code		1		2		3		4	
h_6	0.50	1	0.50	1	0.50	1	0.50	1	0.50	0	
h_3	0.20	00	0.20	00	0.20	00	0.30	00	0.50	1	
h_4	0.15	011	0.15	011	0.20	010	0.20	01			
h_5	0.10	0100	0.10	0100	0.10	011					
h_1	0.07	01010	0.10	0101							
h_2	0.03	01011									

Figure 6.5 Huffman code assignment procedure example.

This entire process is then repeated for each reduced source value until the original source reduction value is reached. The final source code assignment value appears at the bottom left in Figure 6.5. Using Eq. 6.4, the last step in this process is to compute the average length of the code.

$$L_{avg} = (0.50)(1) + (0.20)(2) + (0.15)(3) + (0.10)(4) + (0.07)(5) + (0.03)(5)$$
$$= 2.25 \text{ bits/symbol}$$

(6.4)

The goal of the Huffman coding procedure is to create optimal code for a set of symbols and probabilities subject to the constraint that the symbols be coded one at a time. After creation of the code, decoding and/or encoding is accomplished in a simple lookup table manner [1]. This allows any string of Huffman encoded symbols to be decoded by inspecting the individual symbols of the string starting from the left and proceeding to the right. As an example, a left-to-right scan of the encoded string 0101100110100 shows that the first code word is 01011, which corresponds to the code for symbol h_2. The next valid code word is 00, which corresponds to the code for symbol h_3. Repeating these steps reveals the entire decoded message to be $h_2 h_3 h_6 h_6 h_5$.

6.5 JPEG2000 decoding

A JPEG2000 decoder simply inverts all of the operations described in the previous sections. After decoding the arithmetically coded coefficients, a user-supplied number of the original image's subbands are then reconstructed [1, 11]. Although the encoder may have arithmetically encoded M_b bit-planes for an individual subband, the user may choose to decode only N_b bit-planes. As a result, this amounts to quantizing the coefficients using a step size of $2^{M_b - N_b} \cdot \Delta_b$. Next, any non-decoded bits are then set to zero and all resulting coefficients, denoted $\overline{q}_b(u,v)$, are denormalized using

$$R_{q_b}(u,v) = \begin{cases} \left(\overline{q}_b(u,v) + 2^{M_b - N_b(u,v)}\right) \cdot \Delta_b & \overline{q}_b(u,v) > 0 \\ \left(\overline{q}_b(u,v) - 2^{M_b - N_b(u,v)}\right) \cdot \Delta_b & \overline{q}_b(u,v) < 0 \\ 0 & \overline{q}_b(u,v) < 0 \end{cases}$$

(6.5)

where $R_{q_b}(u,v)$ denotes a denormalized transform coefficient and $N_b(u,v)$ is the number of decoded bit-planes for $\overline{q}_b(u,v)$. The denormalized coefficients are then inverse transformed and level shifted to yield an approximation to the original image. The custom function jpeg2k2im.m, given in Appendix C provided by Gonzalez, et al. [11], is only an approximation to this process and reverses the compression of im2jpeg2k.m described earlier [1, 11].

CHAPTER 7
2-WAY AND 3-WAY HYBRID TEST IMAGE COMPARISONS

7.1 Introduction

The test images chosen for the 3-way hybrid algorithm are listed in Table 7.1. These images were chosen for their combination of various complexities and textures containing smooth and contoured areas. For comparison of the final results from the fractal, JPEG, JPEG 2000, 2-way hybrid and 3-way hybrid algorithms, two measurements were used. These measurements are the CR and PSNR as previously defined by Eqs. 5.1 and 5.2.

Table 7.1 Image dimension sizes in pixels for test images used

Image Name	Image Size			
	64×64	128×128	256×256	512×512
Boy		X	X	
Crowd		X	X	
Fruit	X	X	X	X
Kameraman		X	X	X
Lena		X	X	X
Mosaic		X	X	X
Orchid		X	X	X
Peacock		X	X	X
Peppers		X	X	X
Waterfall		X	X	X

7.2 Results for Boy128.pgm Image

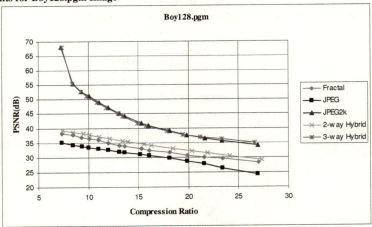

Figure 7.1 Fractal, JPEG, JPEG2000, 2-way & 3-way hybrid results for Boy128.pgm.

The graph shown in Figure 7.1 shows negligible improvement in CR when using the 3-way hybrid algorithm versus the JPEG2000 compression algorithm alone due to disparity in component method CR vs PSNR. However, a great deal of improvement can be realized when using the 3-way hybrid algorithm over the 2-way hybrid algorithm or over the fractal or JPEG algorithm alone. The image itself is an 8-bit grayscale image measuring 128x128 pixels as shown in part (a) of Figure 7.2. Using just the fractal algorithm, the compression ratio ranges from 7.351 to 27.016, the PSNR from 38.376 down to 28.365, with the MSE being varied from 4.99 to 17.46. If just the JPEG compression algorithm is used, the compression ratio varies from 7.351 to 26.928 and the PSNR from 35.202 to 24.259 for quality levels of 68 down to 3. If just the JPEG2000 compression algorithm is used, the compression ratio varies from 7.315 to 26.925 and the PSNR from 68.127 to 34.146 for explicit quantization step sizes of 0.3 up to 7.3. Compressed using the 2-way hybrid algorithm, the compression ratio varies from 7.427 to 27.332 and the PSNR from 39.447 to 29.233. Compressed using the 3-way hybrid algorithm, the compression ratio varies from 7.344 to 26.579 and the PSNR from 68.127 to 35.020. Overall, the average gain in PSNR using the 3-way hybrid algorithm over the 2-way hybrid algorithm is approximately 28.7%. The chart shown in Figure 7.1 was generated from the data listed in Table 7.2.

Figure 7.2 shows examples of the Boy128.pgm image compressed by each method. For this comparison, the image samples were all selected at the data points in Figure 7.1 where CR is approximately 18:1. Part (a) of Figure 7.2 is the original image. Part (b) of Figure 7.2 is the image compressed by fractal with MSE=12.58, CR=18.12, and PSNR=31.759. Part (b) exhibits the "blockiness" typical of fractal methods at high CR [31]. Part (c) of Figure 7.2 is the image compressed by JPEG with Quality=12, CR=18.14, and PSNR=29.794. Part (c) exhibits the loss of detail typical of JPEG at high CR. Part (d) of Figure 7.2 is the image compressed by the 2-way hybrid with CR=18.447 and PSNR=33.12. Part (d) exhibits a subjectively better image than either part (b) or part (c), but its component methods still shows some artifacts, especially in the background. Part (e) of Figure 7.2 is the image compressed by JPEG2000 with Qs=3.9, CR=18.094, and PSNR=39.131. Part (e) shows a lesser degree of loss of detail than part (c) or part (d) but does also show some false contours in smooth areas [31]. Part (f) of Figure 7.2 is the image compressed by the 3-way hybrid with CR=17.844 and PSNR=39.242. Subjectively, part (f) is noticeably better than the other methods and the remaining artifacts are quite small [31].

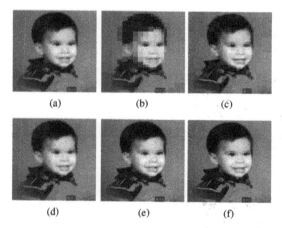

Figure 7.2 Boy128.pgm image at CR approximately 18:1.

Table 7.2 Compression algorithm results for Boy128.pgm image file

Fractal			JPEG			JPEG 2000			2-Way Hybrid		3-Way Hybrid	
MSE	CR	PSNR	Quality	CR	PSNR	Qs	CR	PSNR	CR	PSNR	CR	PSNR
4.99	7.351	38.376	68	7.351	35.202	0.3	7.315	68.127	7.427	39.447	7.344	68.127
5.62	8.444	37.801	60	8.436	34.506	0.5	8.457	55.448	8.427	38.847	8.367	55.448
6.20	9.318	37.137	51	9.349	33.952	0.7	9.322	52.508	9.485	38.318	9.244	52.508
6.78	10.073	36.580	44	10.005	33.590	0.9	10.073	50.969	10.098	37.869	9.855	50.969
7.47	11.028	36.109	36	11.028	33.081	1.2	11.057	48.860	11.103	37.282	10.904	48.860
8.37	11.996	34.985	31	11.996	32.659	1.5	11.961	47.096	12.103	36.621	11.781	47.096
9.04	13.077	34.243	25	13.088	32.076	1.9	13.086	45.095	13.409	35.816	12.933	45.095
9.56	13.620	33.984	23	13.598	31.838	2.1	13.600	44.212	13.945	35.423	13.409	44.216
10.82	15.298	33.008	18	15.198	31.132	2.8	15.298	41.822	15.603	34.656	14.949	41.825
11.13	16.109	32.487	16	16.109	30.736	3.1	16.023	41.004	16.301	34.170	15.799	41.023
12.58	18.120	31.759	12	18.140	29.794	3.9	18.094	39.131	18.447	33.120	17.844	39.242
13.71	19.878	30.710	9	19.878	28.831	4.6	19.841	37.781	20.346	32.257	19.430	38.001
14.88	21.578	30.130	7	21.578	27.911	5.3	21.665	36.677	21.895	31.524	21.215	37.109
16.03	23.394	29.541	5	23.394	26.343	6.0	23.314	35.682	24.152	30.506	23.327	36.278
17.46	27.016	28.365	3	26.928	24.259	7.3	26.925	34.146	27.332	29.233	26.579	35.020

7.3 Results for Boy256.pgm Image

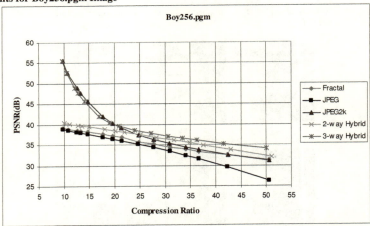

Figure 7.3 Fractal, JPEG, JPEG2000, 2-way & 3-way hybrid results for Boy256.pgm.

The graph shown in Figure 7.3 shows the improvement in CR when using the 3-way hybrid algorithm versus the fractal, JPEG, JPEG2000, or 2-way hybrid compression algorithm alone. The image itself is an 8-bit grayscale image measuring 256x256 pixels as shown in part (a) of Figure 7.4. Using just the fractal algorithm, the compression ratio ranges from 9.793 to 50.579, the PSNR from 39.132 down to 31.339, with the MSE being varied from 4.47 to 12.62. If just the JPEG compression algorithm is used, the compression ratio varies from 9.817 to 50.619 and the PSNR from 38.935 to 26.165 for quality levels of 76 down to 3. If just the JPEG2000 compression algorithm is used, the compression ratio varies from 9.815 to 50.607 and the PSNR from 55.551 to 31.148 for explicit quantization step sizes of 0.5 up to 10.9. Compressed using the 2-way hybrid algorithm, the compression ratio varies from 10.069 to 51.132 and the PSNR from 40.349 to 32.080. Compressed using the 3-way hybrid algorithm, the compression ratio varies from 9.683 to 50.115 and the PSNR from 55.551 to 34.055. Overall, the average gain in PSNR using the 3-way hybrid algorithm over the 2-way hybrid algorithm is approximately 11.3%. The chart shown in Figure 7.3 was generated from the data listed in Table 7.3.

Figure 7.4 shows examples of the Boy256.pgm image compressed by each method. For this comparison, the image samples were all selected at the data points in Figure 7.3 where CR is approximately 36:1. Part (a) of Figure 7.4 is the original image. Part (b) of Figure 7.4 is the image compressed by fractal with MSE=10.07, CR=36.662, and PSNR=33.343. The fractal image shows significant blockiness. Part (c) of Figure 7.4 is the image compressed by JPEG with Quality=9, CR=36.682, and PSNR=31.605. Part (c) exhibits the loss of detail typical of JPEG at high CR. Part (d) of Figure 7.4 is the image compressed by the 2-way hybrid with CR=37.544 and PSNR=34.922. Part (e) of Figure 7.4 is the image compressed by JPEG2000 with Qs=7.9, CR=36.674, and PSNR=33.805. Part (f) of Figure 7.4 is the image compressed by the 3-way hybrid with CR=36.276 and PSNR=36.135. The resulting 3-way hybrid image of part (f) is better than the other methods.

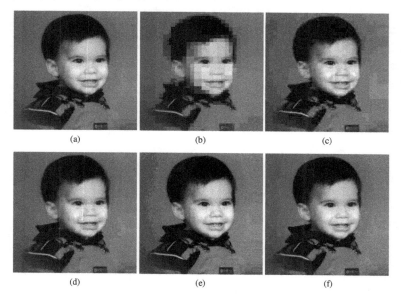

(a) (b) (c)

(d) (e) (f)

Figure 7.4 Boy256.pgm image at CR approximately 36:1.

Table 7.3 Compression algorithm results for Boy256.pgm image file

Fractal			JPEG			JPEG 2000			2-Way Hybrid		3-Way Hybrid	
MSE	CR	PSNR	Quality	CR	PSNR	Qs	CR	PSNR	CR	PSNR	CR	PSNR
4.47	9.793	39.132	76	9.817	38.935	0.5	9.815	55.551	10.069	40.349	9.683	55.551
4.58	10.783	38.924	72	10.873	38.567	0.7	10.746	52.624	11.124	40.149	10.616	52.624
4.77	12.625	38.539	66	12.553	38.103	1.2	12.642	48.916	12.810	39.832	12.094	48.919
4.86	13.245	38.404	63	13.221	37.926	1.4	13.331	47.655	13.496	39.702	12.728	47.656
5.03	14.711	38.198	57	14.721	37.608	1.8	14.690	45.589	14.993	39.470	14.040	45.590
5.57	17.664	37.683	43	17.664	36.897	2.8	17.698	41.962	18.083	38.905	16.864	41.990
5.98	19.632	37.192	35	19.620	36.389	3.4	19.654	40.235	20.108	38.488	19.017	40.415
6.41	21.352	36.842	30	21.359	35.955	3.9	21.335	39.238	21.865	38.109	20.936	39.636
7.37	24.736	35.668	23	24.680	35.168	4.9	24.759	37.409	25.144	37.445	24.038	38.537
8.04	27.741	35.153	18	27.717	34.380	5.7	27.752	36.221	28.279	36.784	27.279	37.756
8.79	30.906	34.400	14	30.949	33.449	6.5	30.891	35.259	31.682	36.083	30.531	37.094
9.50	34.070	33.999	11	34.052	32.469	7.3	34.058	34.340	34.886	35.455	33.427	36.572
10.07	36.662	33.343	9	36.682	31.605	7.9	36.674	33.805	37.544	34.922	36.276	36.135
11.08	42.346	32.555	6	42.346	29.637	9.2	42.411	32.540	42.956	33.767	41.567	35.200
12.62	50.579	31.339	3	50.619	26.165	10.9	50.607	31.148	51.132	32.080	50.115	34.055

7.4 Results for Crowd128.pgm Image

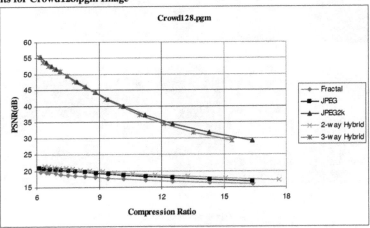

Figure 7.5 Fractal, JPEG, JPEG2000, 2-way & 3-way hybrid results for Crowd128.pgm.

The graph shown in Figure 7.5 shows negligible improvement in CR when using the 3-way hybrid algorithm versus the JPEG2000 compression algorithm alone. However, a great deal of improvement can be realized when using the 3-way hybrid algorithm over the 2-way hybrid algorithm or over the fractal or JPEG algorithms alone. The image itself is an 8-bit grayscale image measuring 128x128 pixels as shown in part (a) of Figure 7.6. Using just the fractal algorithm, the compression ratio ranges from 6.184 to 16.383, the PSNR from 19.760 down to 16.029, with the MSE being varied from 35.66 to 49.61. If just the JPEG compression algorithm is used, the compression ratio varies from 6.133 to 16.366 and the PSNR from 20.874 to 16.678 for quality levels of 17 down to 3. If just the JPEG2000 compression algorithm is used, the compression ratio varies from 6.184 to 16.376 and the PSNR from 55.537 to 29.134 for explicit quantization step sizes of 0.5 up to 12.4. Compressed using the 2-way

hybrid algorithm, the compression ratio varies from 6.376 to 17.235 and the PSNR from 21.470 to 17.235. Compressed using the 3-way hybrid algorithm, the compression ratio varies from 6.128 to 15.384 and the PSNR from 55.537 to 29.134. Overall, the average gain in PSNR using the 3-way hybrid algorithm over the 2-way hybrid algorithm is approximately 122.3%. The chart shown in Figure 7.5 was generated from the data listed in Table 7.4.

Figure 7.6 shows examples of the Crowd128.pgm image compressed by each method. For this comparison, the image samples were all selected at the data points in Figure 7.5 where CR is approximately 11:1. Part (a) of Figure 7.6 is the original image. Part (b) of Figure 7.6 is the image compressed by fractal with MSE=45.28, CR=11.232, and PSNR=17.173. Part (b) shows evidence of the "blockiness" typical of fractal methods at high CR. Part (c) of Figure 7.6 is the image compressed by JPEG with Quality=6, CR=11.232, and PSNR=18.386. Part (d) of Figure 7.6 is the image compressed by the 2-way hybrid with CR=11.697 and PSNR=37.198. Part (e) of Figure 7.6 is the image compressed by JPEG2000 with Qs=4.8, CR=11.218, and PSNR=37.198. Part (f) of Figure 7.6 is the image compressed by the 3-way hybrid with CR=10.977 and PSNR=37.198. Subjectively, the 3-way hybrid of part (f) is better than the other methods.

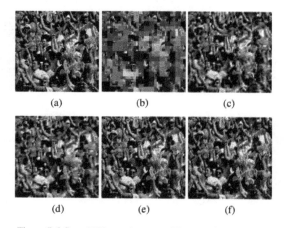

(a)   (b)   (c)

(d)   (e)   (f)

Figure 7.6 Crowd128.pgm image at CR approximately 11:1.

Table 7.4 Compression algorithm results for Crowd128.pgm image file

Fractal			JPEG			JPEG 2000			2-Way Hybrid		3-Way Hybrid	
MSE	CR	PSNR	Quality	CR	PSNR	Qs	CR	PSNR	CR	PSNR	CR	PSNR
35.66	6.184	19.760	17	6.133	20.874	0.5	6.184	55.537	6.376	21.470	6.128	55.537
36.58	6.461	19.559	16	6.354	20.721	0.6	6.463	53.687	6.575	21.310	6.329	53.687
37.01	6.612	19.432	15	6.615	20.560	0.7	6.707	52.583	6.847	21.153	6.578	52.583
37.85	6.896	19.234	14	6.896	20.407	0.8	6.928	51.830	7.111	20.992	6.862	51.830
38.42	7.152	19.025	13	7.171	20.246	0.9	7.132	51.035	7.417	20.805	7.139	51.035
39.25	7.522	18.759	12	7.533	20.066	1.1	7.489	49.550	7.739	20.603	7.488	49.550
39.91	7.869	18.603	11	7.869	19.876	1.4	7.934	47.675	8.106	20.351	7.809	47.675
40.72	8.333	18.302	10	8.324	19.660	1.7	8.342	46.084	8.555	20.099	8.291	46.084
41.84	8.826	18.055	9	8.850	19.416	2.1	8.820	44.346	9.177	19.853	8.788	44.346
42.77	9.414	17.829	8	9.441	19.126	2.7	9.424	42.195	9.744	19.559	9.339	42.195
43.79	10.173	17.572	7	10.167	18.789	3.5	10.169	39.983	10.607	19.209	10.055	39.983
45.28	11.232	17.173	6	11.232	18.386	4.8	11.218	37.198	11.697	18.815	10.977	37.198
46.80	12.547	16.781	5	12.528	17.925	6.6	12.543	34.421	13.257	18.390	12.120	34.421
48.24	14.310	16.448	4	14.310	17.361	9.2	14.306	31.640	15.073	17.835	13.531	31.640
49.61	16.383	16.029	3	16.366	16.678	12.4	16.376	29.134	17.652	17.235	15.384	29.134

7.5 Results for Crowd256.pgm Image

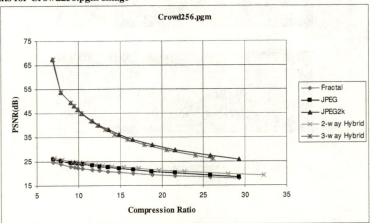

Figure 7.7 Fractal, JPEG, JPEG2000, 2-way & 3-way hybrid results for Crowd256.pgm.

The graph shown in Figure 7.7 shows negligible improvement in CR when using the 3-way hybrid algorithm versus the JPEG2000 compression algorithm alone. However, a great deal of improvement can be realized when using the 3-way hybrid algorithm over the 2-way hybrid algorithm or over the fractal or JPEG algorithm alone. The image itself is an 8-bit grayscale image measuring 256x256 pixels as shown in part (a) of Figure 7.8. Using just the fractal algorithm, the compression ratio ranges from 6.922 to 29.264, the PSNR from 24.719 down to 17.840, with the MSE being varied from 21.67 to 42.52. If just the JPEG compression algorithm is used, the compression ratio varies from 6.922 to 29.264 and the PSNR from 26.059 to 18.507 for quality levels of 24 down to 2. If just the JPEG2000 compression algorithm is used, the compression ratio varies from 6.909 to 29.280 and the PSNR from 67.487 to 25.913 for explicit quantization step sizes of 0.3 up to 18.9. Compressed using the 2-way

hybrid algorithm, the compression ratio varies from 7.085 to 32.259 and the PSNR from 26.675 to 19.249. Compressed using the 3-way hybrid algorithm, the compression ratio varies from 6.926 to 26.158 and the PSNR from 67.487 to 25.951. Overall, the average gain in PSNR using the 3-way hybrid algorithm over the 2-way hybrid algorithm is approximately 74.6%. The chart shown in Figure 7.7 was generated from the data listed in Table 7.5.

Figure 7.8 shows examples of the Crowd256.pgm image compressed by each method. For this comparison, the image samples were all selected at the data points in Figure 7.7 where CR is approximately 12:1. Part (a) of Figure 7.8 is the original image. Part (b) of Figure 7.8 is the image compressed by fractal with MSE=30.92, CR=12.488, and PSNR=21.246. Part (c) of Figure 7.8 is the image compressed by JPEG with Quality=9, CR=12.495, and PSNR=22.935. Part (d) of Figure 7.8 is the image compressed by the 2-way hybrid with CR=12.891 and PSNR=23.411. The 2-way hybrid eliminates the majority of the problems seen in fractal or JPEG alone. Part (e) of Figure 7.8 is the image compressed by JPEG2000 with Qs=3.4, CR=12.474, and PSNR=40.099. Part (f) of Figure 7.8 is the image compressed by the 3-way hybrid with CR=12.269 and PSNR=40.099. Subjectively, the 3-way hybrid of part (f) is better than the other methods.

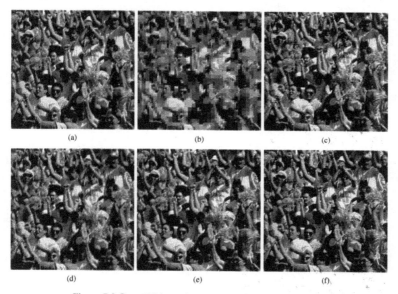

Figure 7.8 Crowd256.pgm image at CR approximately 12:1.

Table 7.5 Compression algorithm results for Crowd256.pgm image file

Fractal			JPEG			JPEG 2000			2-Way Hybrid		3-Way Hybrid	
MSE	CR	PSNR	Quality	CR	PSNR	Qs	CR	PSNR	CR	PSNR	CR	PSNR
21.67	6.922	24.719	24	6.922	26.059	0.3	6.909	67.487	7.085	26.675	6.926	67.487
23.85	7.919	23.820	19	7.919	25.324	0.6	7.937	53.707	8.085	25.860	7.899	53.707
26.24	9.095	22.918	15	9.099	24.589	1.1	9.085	49.528	9.332	25.061	9.038	49.528
26.92	9.531	22.614	14	9.535	24.373	1.3	9.450	48.197	9.763	24.842	9.436	48.197
27.47	9.930	22.391	13	9.924	24.162	1.6	9.961	46.471	10.195	24.624	9.831	46.471
28.28	10.480	22.082	12	10.483	23.890	1.9	10.442	45.068	10.797	24.353	10.347	45.068
29.84	11.676	21.546	10	11.676	23.293	2.8	11.681	41.796	12.028	23.764	11.506	41.796
30.92	12.488	21.246	9	12.495	22.935	3.4	12.474	40.099	12.891	23.411	12.269	40.099
31.90	13.544	20.826	8	13.535	22.518	4.3	13.583	38.123	13.986	22.980	13.195	38.123
33.23	14.837	20.455	7	14.831	22.058	5.4	14.850	36.157	15.406	22.516	14.353	36.158
34.84	16.574	20.042	6	16.578	21.530	7.0	16.533	33.990	17.241	22.012	15.891	33.992
36.81	18.836	19.448	5	18.820	20.903	9.1	18.840	31.817	19.632	21.426	17.905	31.826
38.56	21.570	18.926	4	21.570	20.139	11.6	21.608	29.815	22.729	20.721	20.549	29.827
40.96	25.869	18.297	3	25.858	19.167	15.7	25.860	27.364	27.965	19.840	24.020	27.397
42.52	29.264	17.840	2	29.264	18.507	18.9	29.280	25.913	32.259	19.249	26.158	25.951

7.6 Results for Fruit64.pgm Image

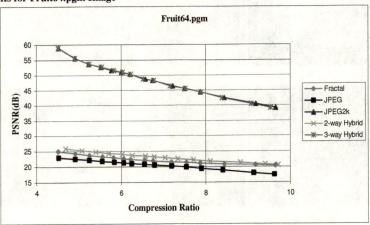

Figure 7.9 Fractal, JPEG, JPEG2000, 2-way & 3-way hybrid results for Fruit64.pgm.

The graph shown in Figure 7.9 shows negligible improvement in CR when using the 3-way hybrid algorithm versus the JPEG2000 compression algorithm alone. The image itself is an 8-bit grayscale image measuring 64x64 pixels as shown in part (a) of Figure 7.10. Using just the fractal algorithm, the compression ratio ranges from 4.515 to 9.646, the PSNR from 24.969 down to 20.426, with the MSE being varied from 22.71 to 35.11. If just the JPEG compression algorithm is used, the compression ratio varies from 4.525 to 9.623 and the PSNR from 22.937 to 17.495 for quality levels of 23 down to 2. If just the JPEG2000 compression algorithm is used, the compression ratio varies from 4.522 to 9.655 and the PSNR from 58.914 to 39.124 for explicit quantization step sizes of 0.4 up to 3.8. Compressed using the 2-way hybrid algorithm, the compression ratio varies from 4.691 to 9.714 and the PSNR from 25.844 to 20.574. Compressed using the 3-way hybrid algorithm, the compression ratio varies from 4.520 to 9.534 and the PSNR from 58.914 to 39.124. Overall, the average gain in PSNR using the 3-way hybrid algorithm over the 2-way hybrid algorithm is approximately 109.8%. The chart shown in Figure 7.9 was generated from the data listed in Table 7.6.

Figure 7.10 shows examples of the Fruit64.pgm image compressed by each method. For this comparison, the image samples were all selected at the data points in Figure 7.9 where CR is approximately 7:1. Part (a) of Figure 7.10 is the original image. Part (b) of Figure 7.10 is the image compressed by fractal with MSE=30.93, CR=7.184, and PSNR=21.508. Part (c) of Figure 7.10 is the image compressed by JPEG with Quality=7, CR=7.196, and PSNR=20.248. Part (d) of Figure 7.10 is the image compressed by the 2-way hybrid with CR=7.338 and PSNR=22.511. The 2-way hybrid eliminates the majority of the problems seen in fractal or JPEG alone. Part (e) of Figure 7.10 is the image compressed by JPEG2000 with Qs=1.6, CR=7.218, and PSNR=46.392. Part (f) of Figure 7.10 is the image compressed by the 3-way hybrid with CR=7.159 and PSNR=46.392. For this image, with the exception of the blockiness generated by fractal, none of the other methods showed excessive distortion. Subjectively, the 3-way hybrid of part (f) is better than the other methods.

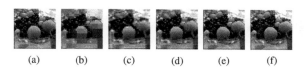

(a)   (b)   (c)   (d)   (e)   (f)

Figure 7.10 Fruit64.pgm image at CR approximately 7:1.

Table 7.6 Compression algorithm results for Fruit64.pgm image file

Fractal			JPEG			JPEG 2000			2-Way Hybrid		3-Way Hybrid	
MSE	CR	PSNR	Quality	CR	PSNR	Qs	CR	PSNR	CR	PSNR	CR	PSNR
22.71	4.515	24.969	23	4.525	22.937	0.4	4.522	58.914	4.691	25.844	4.520	58.914
24.37	4.915	24.236	19	4.915	22.524	0.5	4.913	55.527	5.073	25.204	4.903	55.527
25.48	5.254	23.692	16	5.248	22.122	0.6	5.236	53.636	5.399	24.747	5.241	53.636
26.17	5.560	23.247	14	5.553	21.803	0.7	5.522	52.698	5.707	24.359	5.538	52.698
27.18	5.820	22.956	12	5.837	21.472	0.8	5.777	51.613	6.025	24.012	5.828	51.613
28.10	6.060	22.617	11	6.060	21.277	0.9	6.006	51.049	6.264	23.655	6.034	51.049
28.47	6.226	22.366	10	6.254	21.033	1.0	6.213	50.402	6.574	23.438	6.226	50.402
29.45	6.564	22.050	9	6.553	20.820	1.2	6.575	48.747	6.781	23.169	6.522	48.747
30.02	6.758	21.866	8	6.792	20.513	1.3	6.745	48.228	7.036	22.886	6.769	48.228
30.93	7.184	21.508	7	7.196	20.248	1.6	7.218	46.392	7.338	22.511	7.159	46.392
31.83	7.539	21.348	6	7.539	19.870	1.8	7.498	45.508	7.709	22.120	7.512	45.508
32.42	7.872	21.092	5	7.902	19.439	2.1	7.873	44.320	8.105	21.788	7.887	44.320
33.67	8.437	20.862	4	8.403	18.924	2.6	8.437	42.431	8.818	21.337	8.386	42.431
34.56	9.172	20.624	3	9.131	18.015	3.3	9.184	40.452	9.403	20.846	9.071	40.452
35.11	9.646	20.426	2	9.623	17.495	3.8	9.655	39.124	9.714	20.574	9.534	39.124

7.7 Results for Fruit128.pgm Image

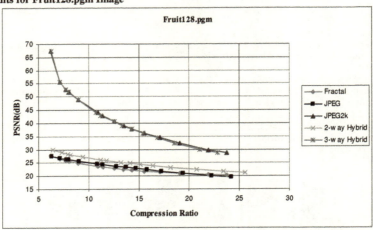

Figure 7.11 Fractal, JPEG, JPEG2000, 2-way & 3-way hybrid results for Fruit128.pgm.

The graph shown in Figure 7.11 shows negligible improvement in CR when using the 3-way hybrid algorithm versus the JPEG2000 compression algorithm alone. However, a great deal of improvement can be realized when using the 3-way hybrid algorithm over the 2-way hybrid algorithm or over the fractal or JPEG algorithm alone. The image itself is an 8-bit grayscale image measuring 128x128 pixels as shown in part (a) of Figure 7.12. Using just the fractal algorithm, the compression ratio ranges from 6.231 to 23.732, the PSNR from 27.678 down to 20.349, with the MSE being varied from 16.84 to 35.15. If just the JPEG compression algorithm is used, the compression ratio varies from 6.278 to 24.187 and the PSNR from 27.470 to 19.509 for quality levels of 34 down to 2. If just the JPEG2000 compression algorithm is used, the compression ratio varies from 6.231 to 23.762 and the PSNR from 67.375 to 28.632 for explicit quantization step sizes of 0.3 up to 14.2. Compressed using the 2-way hybrid algorithm, the compression ratio varies from 6.434 to 25.583 and the PSNR from 29.932 to 21.232. Compressed using the 3-way hybrid algorithm, the compression ratio varies from 6.286 to 22.808 and the PSNR from 67.375 to 28.706. Overall, the average gain in PSNR using the 3-way hybrid algorithm over the 2-way hybrid algorithm is approximately 66%. The chart shown in Figure 7.11 was generated from the data listed in Table 7.7.

Figure 7.12 shows examples of the Fruit128.pgm image compressed by each method. For this comparison, the image samples were all selected at the data points in Figure 7.11 where CR is approximately 13:1. Part (a) of Figure 7.12 is the original image. Part (b) of Figure 7.12 is the image compressed by fractal with MSE=27.12, CR=12.547, and PSNR=22.844. Part (c) of Figure 7.12 is the image compressed by JPEG with Quality=9, CR=12.762, and PSNR=23.719. Part (d) of Figure 7.12 is the image compressed by the 2-way hybrid with CR=13.182 and PSNR=25.192. Part (e) of Figure 7.12 is the image compressed by JPEG2000 with Qs=3.3, CR=12.548, and PSNR=40.63. Part (f) of Figure 7.12 is the image compressed by the 3-way hybrid with CR=12.557 and PSNR=40.634. For this image, with the exception of the blockiness generated by fractal, none of the other methods showed excessive distortion. Subjectively, the 3-way hybrid of part (f) is better than the other methods.

106

(a)   (b)   (c)

(d)   (e)   (f)

Figure 7.12 Fruit128.pgm image at CR approximately 13:1.

Table 7.7 Compression algorithm results for Fruit128.pgm image file

Fractal			JPEG			JPEG 2000			2-Way Hybrid		3-Way Hybrid	
MSE	CR	PSNR	Quality	CR	PSNR	Qs	CR	PSNR	CR	PSNR	CR	PSNR
16.84	6.231	27.678	34	6.278	27.470	0.3	6.231	67.375	6.434	29.932	6.286	67.375
18.75	7.136	26.439	27	7.130	26.819	0.5	7.116	55.531	7.337	28.895	7.127	55.531
20.04	7.761	25.790	23	7.779	26.387	0.7	7.764	52.606	7.934	28.281	7.750	52.606
20.61	8.090	25.479	22	8.007	26.275	0.8	8.042	51.738	8.171	28.034	7.984	51.738
22.33	8.996	24.795	17	9.065	25.615	1.2	8.998	48.868	9.398	27.273	8.996	48.870
25.19	10.896	23.737	12	10.875	24.581	2.2	10.892	43.982	11.171	26.112	10.732	43.982
25.82	11.420	23.476	11	11.380	24.357	2.5	11.370	42.907	11.798	25.848	11.194	42.907
27.12	12.547	22.844	9	12.762	23.719	3.3	12.548	40.630	13.182	25.192	12.557	40.634
28.65	13.486	22.406	8	13.746	23.309	4.0	13.521	38.903	14.113	24.756	13.343	38.908
29.50	14.297	22.112	7	14.694	22.900	4.6	14.306	37.863	15.156	24.361	14.335	37.863
30.41	15.544	21.831	6	15.799	22.451	5.6	15.545	36.241	16.399	23.897	15.427	36.196
31.50	17.100	21.452	5	17.226	21.773	7.0	17.116	34.444	18.181	23.257	16.871	34.450
32.74	19.069	21.086	4	19.407	21.012	9.0	19.079	32.240	20.758	22.520	18.530	32.253
34.40	21.924	20.531	3	22.221	20.056	12.2	21.933	29.748	23.394	21.697	21.051	29.789
35.15	23.732	20.349	2	24.187	19.509	14.2	23.762	28.632	25.583	21.232	22.808	28.706

7.8 Results for Fruit256.pgm Image

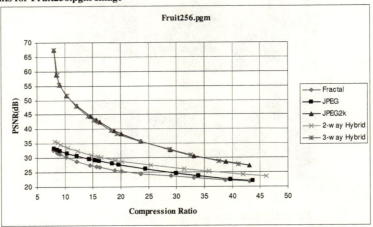

Figure 7.13 Fractal, JPEG, JPEG2000, 2-way & 3-way hybrid results for Fruit256.pgm.

The graph shown in Figure 7.13 shows negligible improvement in CR when using the 3-way hybrid algorithm versus the JPEG2000 compression algorithm alone. However, a great deal of improvement can be realized when using the 3-way hybrid algorithm over the 2-way hybrid algorithm or over the fractal or JPEG algorithm alone. The image itself is an 8-bit grayscale image measuring 256x256 pixels as shown in part (a) of Figure 7.14. Using just the fractal algorithm, the compression ratio ranges from 7.967 to 43.069, the PSNR from 32.639 down to 21.739, with the MSE being varied from 9.50 to 30.26. If just the JPEG compression algorithm is used, the compression ratio varies from 7.998 to 43.701 and the PSNR from 33.294 to 21.912 for quality levels of 48 down to 2. If just the JPEG2000 compression algorithm is used, the compression ratio varies from 7.996 to 43.066 and the PSNR from 67.482 to 27.151 for explicit quantization step sizes of 0.3 up to 18.1. Compressed using the 2-way hybrid algorithm, the compression ratio varies from 8.109 to 46.098 and the PSNR from 35.805 to 23.558. Compressed using the 3-way hybrid algorithm, the compression ratio varies from 7.996 to 41.124 and the PSNR from 67.482 to 27.634. Overall, the average gain in PSNR using the 3-way hybrid algorithm over the 2-way hybrid algorithm is approximately 41.8%. The chart shown in Figure 7.13 was generated from the data listed in Table 7.8.

Figure 7.14 shows examples of the Fruit256.pgm image compressed by each method. For this comparison, the image samples were all selected at the data points in Figure 7.13 where CR is approximately 20:1. Part (a) of Figure 7.14 is the original image. Part (b) of Figure 7.14 is the image compressed by fractal with MSE=20.89, CR=20.102, and PSNR=25.399. Part (c) of Figure 7.14 is the image compressed by JPEG with Quality=10, CR=19.614, and PSNR=27.605. Part (d) of Figure 7.14 is the image compressed by the 2-way hybrid with CR=20.219 and PSNR=28.804. Part (e) of Figure 7.14 is the image compressed by JPEG2000 with Qs=4.5, CR=20.083, and PSNR=38.384. Part (f) of Figure 7.14 is the image compressed by the 3-way hybrid with CR=19.274 and PSNR=38.405. For this image, with the exception of the blockiness generated by fractal, none of the other methods showed excessive distortion. Subjectively, the 3-way hybrid of part (f) is better than the other methods.

(a)   (b)   (c)

(d)   (e)   (f)

Figure 7.14 Fruit256.pgm image at CR approximately 20:1.

Table 7.8 Compression algorithm results for Fruit256.pgm image file

Fractal			JPEG			JPEG 2000			2-Way Hybrid		3-Way Hybrid	
MSE	CR	PSNR	Quality	CR	PSNR	Qs	CR	PSNR	CR	PSNR	CR	PSNR
9.50	7.967	32.639	48	7.998	33.294	0.3	7.996	67.482	8.109	35.805	7.996	67.482
10.30	8.605	31.893	42	8.625	32.796	0.4	8.572	58.936	8.746	35.187	8.609	58.936
11.00	9.113	31.348	38	9.078	32.481	0.5	9.053	55.360	9.176	34.712	9.028	55.362
12.47	10.323	30.225	30	10.382	31.617	0.8	10.330	51.726	10.524	33.572	10.336	51.737
14.57	12.094	28.796	23	12.097	30.710	1.3	12.101	48.251	12.294	32.352	11.977	48.020
16.70	14.518	27.519	17	14.407	29.629	2.1	14.570	44.520	14.704	31.011	14.195	44.433
17.67	15.656	27.049	15	15.456	29.179	2.5	15.674	43.104	15.845	30.495	15.241	43.121
18.11	16.274	26.841	14	16.122	28.910	2.7	16.197	42.500	16.499	30.221	15.853	42.519
20.04	18.929	25.647	11	18.496	27.976	3.9	18.880	39.521	19.089	29.201	18.219	39.516
20.89	20.102	25.399	10	19.614	27.605	4.5	20.083	38.384	20.219	28.804	19.274	38.405
23.01	23.639	24.462	7	24.487	26.117	6.3	23.600	35.695	25.417	27.434	23.554	35.765
25.57	28.966	23.607	5	29.973	24.656	9.3	28.893	32.593	31.409	26.050	28.801	32.784
27.09	33.107	22.961	4	33.999	23.642	11.9	33.183	30.568	35.840	25.161	32.596	30.808
29.18	38.765	22.220	3	39.656	22.503	15.5	38.865	28.451	41.939	24.124	37.565	28.794
30.26	43.069	21.739	2	43.701	21.912	18.1	43.066	27.151	46.098	23.558	41.124	27.634

7.9 Results for Fruit512.pgm Image

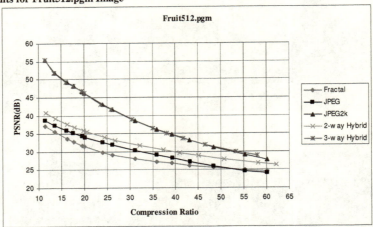

Figure 7.15 Fractal, JPEG, JPEG2000, 2-way & 3-way hybrid results for Fruit512.pgm.

The graph shown in Figure 7.15 shows negligible improvement in CR when using the 3-way hybrid algorithm versus the JPEG2000 compression algorithm alone. However, a great deal of improvement can be realized when using the 3-way hybrid algorithm over the 2-way hybrid algorithm or over the fractal or JPEG algorithm alone. The image itself is an 8-bit grayscale image measuring 512x512 pixels as shown in Figure 7.16. Using just the fractal algorithm, the compression ratio ranges from 11.476 to 60.101, the PSNR from 37.083 down to 24.528, with the MSE being varied from 6.12 to 23.04. If just the JPEG compression algorithm is used, the compression ratio varies from 11.479 to 60.128 and the PSNR from 38.723 to 23.968 for quality levels of 52 down to 2. If just the JPEG2000 compression algorithm is used, the compression ratio varies from 11.465 to 60.059 and the PSNR from 55.433 to 27.712 for explicit quantization step sizes of 0.5 up to 17.7. Compressed using the 2-way hybrid algorithm, the compression ratio varies from 11.563 to 62.123 and the PSNR from 40.722 to 26.351. Compressed using the 3-way hybrid algorithm, the compression ratio varies from 11.435 to 58.013 and the PSNR from 55.413 to 28.986. Overall, the average gain in PSNR using the 3-way hybrid algorithm over the 2-way hybrid algorithm is approximately 23.6%. The chart shown in Figure 7.15 was generated from the data listed in Table 7.9.

For this comparison, the image samples were all selected at the data points in Figure 7.15 where CR is approximately 31:1. Figure 7.16 is the original image. Figure 7.17 is the image compressed by fractal with MSE=16, CR=31.18, and PSNR=28.052. Figure 7.18 is the image compressed by JPEG with Quality=9, CR=31.165, and PSNR=30.313. Figure 7.19 is the image compressed by the 2-way hybrid with CR=32.183 and PSNR=31.637. Figure 7.20 is the image compressed by JPEG2000 with Qs=4.8, CR=31.202, and PSNR=38.594. Figure 7.21 is the image compressed by the 3-way hybrid with CR=30.523 and PSNR=38.811. For this image, with the exception of the blockiness generated by fractal, none of the other methods showed excessive distortion. Subjectively, the 3-way hybrid of Figure 7.21 is better than the other methods.

Table 7.9 Compression algorithm results for Fruit512.pgm image file

Fractal			JPEG			JPEG 2000			2-Way Hybrid		3-Way Hybrid	
MSE	CR	PSNR	Quality	CR	PSNR	Qs	CR	PSNR	CR	PSNR	CR	PSNR
6.12	11.476	37.083	52	11.479	38.723	0.5	11.465	55.433	11.563	40.722	11.435	55.413
7.27	13.537	35.509	38	13.564	37.306	0.8	13.541	51.814	13.683	39.194	13.492	51.835
8.89	16.317	33.635	28	16.160	35.888	1.2	16.316	49.169	16.333	37.614	16.054	49.188
9.61	17.668	32.755	24	17.662	35.183	1.4	17.719	48.073	17.884	36.854	17.535	48.092
10.67	19.577	31.721	20	19.582	34.305	1.7	19.549	46.627	19.913	35.873	19.415	46.655
10.95	20.087	31.530	19	20.177	34.050	1.8	20.082	46.238	20.533	35.618	19.985	46.251
13.02	24.047	29.864	14	23.987	32.581	2.7	24.050	43.136	24.451	33.968	23.616	43.127
13.95	26.219	29.141	12	26.219	31.808	3.3	26.240	41.583	26.808	33.147	25.839	41.674
16.00	31.180	28.052	9	31.165	30.313	4.8	31.202	38.594	32.183	31.637	30.523	38.811
17.37	35.981	27.104	7	35.991	29.039	6.5	35.948	36.145	37.345	30.429	35.081	36.345
18.27	39.198	26.737	6	39.298	28.174	7.8	39.193	34.627	40.982	29.689	38.099	35.011
19.35	43.140	26.097	5	43.175	27.215	9.4	43.089	33.054	45.091	28.857	41.692	33.638
20.63	48.387	25.513	4	48.333	25.941	11.9	48.429	31.115	50.669	27.887	46.466	31.804
22.08	55.238	24.837	3	55.250	24.540	15.3	55.235	29.018	58.154	26.864	53.317	29.903
23.04	60.101	24.528	2	60.128	23.968	17.7	60.059	27.712	62.123	26.351	58.013	28.986

Figure 7.16 Original Fruit512.pgm image.

Figure 7.17 Fruit512.pgm image using fractal $(MSE = 16.00, CR = 31.180, PSNR = 28.052)$.

Figure 7.18 Fruit512.pgm image using JPEG $(Quality = 9, CR = 31.165, PSNR = 30.313)$.

Figure 7.19 Fruit512.pgm image using 2-way hybrid $(CR = 32.183, PSNR = 31.637)$.

Figure 7.20 Fruit512.pgm image using JPEG2000 $(Qs = 4.8, CR = 31.202, PSNR = 38.594)$.

Figure 7.21 Fruit512.pgm image using 3-way hybrid $(CR = 30.523, PSNR = 38.811)$.

7.10 Results for Kameraman128.pgm Image

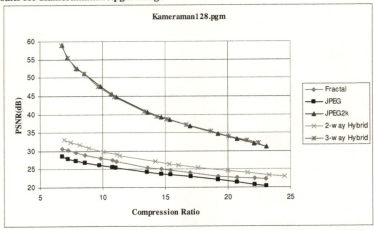

Figure 7.22 Fractal, JPEG, JPEG2000, 2-way & 3-way hybrid results for Kameraman128.pgm.

The graph shown in Figure 7.22 shows negligible improvement in CR when using the 3-way hybrid algorithm versus the JPEG2000 compression algorithm alone. However, a great deal of improvement can be realized when using the 3-way hybrid algorithm over the 2-way hybrid algorithm or over the fractal or JPEG algorithm alone. The image itself is an 8-bit grayscale image measuring 128x128 pixels as shown in part (a) of Figure 7.23. Using just the fractal algorithm, the compression ratio ranges from 6.754 to 23.032, the PSNR from 30.726 down to 22.212, with the MSE being varied from 13.33 to 39.25. If just the JPEG compression algorithm is used, the compression ratio varies from 6.754 to 23.097 and the PSNR from 28.508 to 20.287 for quality levels of 45 down to 2. If just the JPEG2000 compression algorithm is used, the compression ratio varies from 6.770 to 23.052 and the PSNR from 58.994 to 31.254 for explicit quantization step sizes of 0.4 up to 10.4. Compressed using the 2-way hybrid algorithm, the compression ratio varies from 6.928 to 24.513 and the PSNR from 33.037 to 23.012. Compressed using the 3-way hybrid algorithm, the compression ratio varies from 6.732 to 22.464 and the PSNR from 58.994 to 32.282. Overall, the average gain in PSNR using the 3-way hybrid algorithm over the 2-way hybrid algorithm is approximately 54.2%. The chart shown in Figure 7.17 was generated from the data listed in Table 7.10.

Figure 7.23 shows examples of the Kameraman128.pgm image compressed by each method. For this comparison, the image samples were all selected at the data points in Figure 7.22 where CR is approximately 15:1. Part (a) of Figure 7.23 is the original image. Part (b) of Figure 7.23 is the image compressed by fractal with MSE=28.54, CR=14.694, and PSNR=25.06. While the foreground kameraman is well rendered by fractal, the low-detail background exhibits excessive blockiness [31]. Part (c) of Figure 7.23 is the image compressed by JPEG with Quality=10, CR=14.708, and PSNR=23.726. Part (d) of Figure 7.23 is the image compressed by the 2-way hybrid with CR=15.298 and PSNR=26.498. Part (e) of Figure 7.23 is the image compressed by JPEG2000 with Qs=3.9, CR=14.694, and PSNR=39.225. Part (f) of Figure 7.23 is the image compressed by the 3-way hybrid with CR=14.347 and PSNR=39.425. Subjectively, the 3-way hybrid of part (f) is better than the other methods.

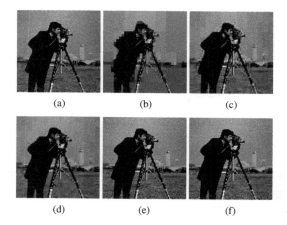

| (a) | (b) | (c) |

| (d) | (e) | (f) |

Figure 7.23 Kameraman128.pgm image at CR approximately 15:1.

Table 7.10 Compression algorithm results for Kameraman128.pgm image file

Fractal			JPEG			JPEG 2000			2-Way Hybrid		3-Way Hybrid	
MSE	CR	PSNR	Quality	CR	PSNR	Qs	CR	PSNR	CR	PSNR	CR	PSNR
13.33	6.754	30.726	45	6.754	28.508	0.4	6.770	58.994	6.928	33.037	6.732	58.994
14.29	7.231	30.283	39	7.234	27.938	0.5	7.211	55.431	7.457	32.415	7.224	55.431
15.68	7.873	29.574	33	7.873	27.298	0.7	7.955	52.569	8.187	31.660	7.869	52.569
17.15	8.577	28.956	29	8.577	26.839	0.9	8.559	51.030	8.888	30.868	8.537	51.030
20.03	9.826	28.019	23	9.727	26.078	1.4	9.824	47.648	10.030	29.810	9.646	47.648
21.51	10.782	27.473	19	10.803	25.532	1.8	10.721	45.549	11.156	29.032	10.642	45.555
22.31	11.110	27.239	18	11.051	25.383	2.0	11.112	44.717	11.380	28.733	10.925	44.720
26.85	13.598	25.499	12	13.553	24.255	3.3	13.600	40.655	14.198	27.085	13.322	40.710
28.54	14.694	25.060	10	14.708	23.726	3.9	14.694	39.225	15.298	26.498	14.347	39.425
29.56	15.413	24.764	9	15.413	23.513	4.3	15.384	38.534	16.046	26.180	15.100	38.705
32.51	16.994	24.001	7	17.011	22.949	5.3	16.965	36.819	17.595	25.485	16.649	37.170
35.85	19.225	23.005	5	19.203	22.076	6.9	19.224	34.699	19.974	24.521	18.635	35.287
36.89	20.706	22.654	4	20.706	21.423	8.2	20.739	33.279	21.578	23.954	20.072	34.056
38.47	22.131	22.425	3	22.131	20.673	9.5	22.103	32.054	23.261	23.345	21.521	32.849
39.25	23.032	22.212	2	23.097	20.287	10.4	23.052	31.254	24.513	23.012	22.464	32.282

7.11 Results for Kameraman256.pgm Image

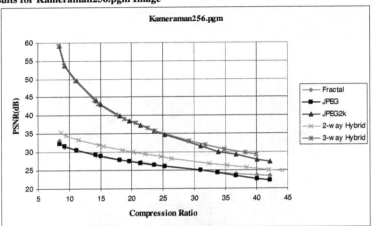

Figure 7.24 Fractal, JPEG, JPEG2000, 2-way & 3-way hybrid results for Kameraman256.pgm.

The graph shown in Figure 7.24 shows negligible improvement in CR when using the 3-way hybrid algorithm versus the JPEG2000 compression algorithm alone until high CR levels are reached. However, a great deal of improvement can be realized when using the 3-way hybrid algorithm over the 2-way hybrid algorithm or over the fractal or JPEG algorithm alone. The image itself is an 8-bit grayscale image measuring 256x256 pixels as shown in part (a) of Figure 7.25. Using just the fractal algorithm, the compression ratio ranges from 8.414 to 42.182, the PSNR from 33.017 down to 23.447, with the MSE being varied from 10.43 to 35.28. If just the JPEG compression algorithm is used, the compression ratio varies from 8.409 to 42.128 and the PSNR from 32.158 to 22.256 for quality levels of 56 down to 2. If just the JPEG2000 compression algorithm is used, the compression ratio varies from 8.414 to 42.179 and the PSNR from 59.111 to 27.221 for explicit quantization step sizes of 0.4 up to 17.7. Compressed using

118

the 2-way hybrid algorithm, the compression ratio varies from 8.627 to 44.172 and the PSNR from 35.353 to 24.787. Compressed using the 3-way hybrid algorithm, the compression ratio varies from 8.380 to 40.019 and the PSNR from 59.111 to 29.220. Overall, the average gain in PSNR using the 3-way hybrid algorithm over the 2-way hybrid algorithm is approximately 32.8%. The chart shown in Figure 7.24 was generated from the data listed in Table 7.11.

Figure 7.25 shows examples of the Kameraman256.pgm image compressed by each method. For this comparison, the image samples were all selected at the data points in Figure 7.24 where CR is approximately 24:1. Part (a) of Figure 7.25 is the original image. Part (b) of Figure 7.25 is the image compressed by fractal with MSE=24.55, CR=23.639, and PSNR=26.644. While the foreground kameraman is well rendered by fractal, the low-detail background exhibits excessive blockiness [31]. Part (c) of Figure 7.25 is the image compressed by JPEG with Quality=10, CR=23.63, and PSNR=26.417. Part (d) of Figure 7.25 is the image compressed by the 2-way hybrid with CR=24.69 and PSNR=28.727. Part (e) of Figure 7.25 is the image compressed by JPEG2000 with Qs=6.3, CR=23.681, and PSNR=35.571. Part (f) of Figure 7.25 is the image compressed by the 3-way hybrid with CR=22.542 and PSNR=36.555. Subjectively, the 3-way hybrid of part (f) is better than the other methods.

(a)   (b)   (c)

(d)   (e)   (f)

Figure 7.25 Kameraman256.pgm image at CR approximately 24:1.

Table 7.11 Compression algorithm results for Kameraman256.pgm image file

Fractal			JPEG			JPEG 2000			2-Way Hybrid		3-Way Hybrid	
MSE	CR	PSNR	Quality	CR	PSNR	Qs	CR	PSNR	CR	PSNR	CR	PSNR
10.43	8.414	33.017	56	8.409	32.158	0.4	8.414	59.111	8.627	35.353	8.380	59.111
11.56	9.303	31.387	48	9.309	31.489	0.6	9.326	53.694	9.542	34.445	9.236	53.694
13.48	11.137	30.497	36	11.142	30.475	1.1	11.134	49.588	11.524	33.236	11.015	49.598
16.41	14.281	29.018	24	14.256	29.159	2.2	14.286	44.083	14.701	31.756	14.001	44.142
17.10	15.062	28.788	22	15.059	28.878	2.5	15.060	43.040	15.522	31.428	14.655	43.057
19.57	18.038	27.916	16	18.038	27.842	3.7	18.069	39.862	18.559	30.378	17.429	40.178
21.37	19.673	27.536	14	19.644	27.451	4.4	19.698	38.458	20.301	29.878	18.874	38.874
22.74	21.429	27.125	12	21.429	27.011	5.2	21.449	37.223	22.191	29.379	20.672	37.927
24.55	23.639	26.644	10	23.630	26.417	6.3	23.681	35.571	24.690	28.727	22.542	36.555
25.89	25.261	26.156	9	25.299	26.136	7.1	25.264	34.534	26.368	28.251	23.793	35.623
29.41	31.023	24.959	6	31.023	24.975	10.7	31.023	31.437	32.419	26.832	29.147	32.859
30.90	33.842	24.420	5	33.842	24.357	12.4	33.869	29.961	35.280	26.293	31.790	31.803
32.55	36.744	23.950	4	36.744	23.539	14.5	36.761	29.213	38.424	25.663	34.812	30.659
34.28	40.092	23.633	3	40.092	22.646	16.7	40.095	27.865	41.939	25.057	38.155	29.664
35.28	42.182	23.447	2	42.128	22.256	17.7	42.179	27.221	44.172	24.787	40.019	29.220

7.12 Results for Kameraman512.pgm Image

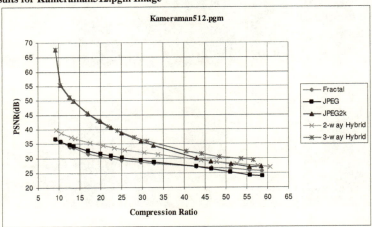

Figure 7.26 Fractal, JPEG, JPEG2000, 2-way & 3-way hybrid results for Kameraman512.pgm.

The graph shown in Figure 7.26 shows negligible improvement in CR when using the 3-way hybrid algorithm versus the JPEG2000 compression algorithm alone until moderate to high CR values are reached. However, a great deal of improvement can be realized when using the 3-way hybrid algorithm over the 2-way hybrid algorithm or over the fractal or JPEG algorithm alone. The image itself is an 8-bit grayscale image measuring 512x512 pixels as shown in Figure 7.27. Using just the fractal algorithm, the compression ratio ranges from 9.202 to 58.518, the PSNR from 36.543 down to 25.757, with the MSE being varied from 6.53 to 27.63. If just the JPEG compression algorithm is used, the compression ratio varies from 9.208 to 58.754 and the PSNR from 36.496 to 23.832 for quality levels of 68 down to 2. If just the JPEG2000 compression algorithm is used, the compression ratio varies from 9.192 to 58.511 and the PSNR from 67.606 to 27.151 for explicit quantization step sizes of 0.3 up to 19.4. Compressed using

120

the 2-way hybrid algorithm, the compression ratio varies from 9.368 to 60.685 and the PSNR from 39.711 to 27.002. Compressed using the 3-way hybrid algorithm, the compression ratio varies from 9.207 to 56.597 and the PSNR from 67.606 to 29.456. Overall, the average gain in PSNR using the 3-way hybrid algorithm over the 2-way hybrid algorithm is approximately 24.1%. The chart shown in Figure 7.26 was generated from the data listed in Table 7.12.

For this comparison, the image samples were all selected at the data points in Figure 7.26 where CR is approximately 33:1. Figure 7.27 is the original image. Figure 7.28 is the image compressed by fractal with MSE=19.05, CR=32.741, and PSNR=28.258. Figure 7.29 is the image compressed by JPEG with Quality=10, CR=32.794, and PSNR=28.718. Figure 7.30 is the image compressed by the 2-way hybrid with CR=33.688 and PSNR=31.218. Figure 7.31 is the image compressed by JPEG2000 with Qs=7.7, CR=32.701, and PSNR=34.296. Figure 7.32 is the image compressed by the 3-way hybrid with CR=31.073 and PSNR=35.804. For this image, with the exception of the blockiness generated by fractal, none of the other methods showed excessive distortion. Subjectively, the 3-way hybrid of Figure 7.32 is better than the other methods.

Table 7.12 Compression algorithm results for Kameraman512.pgm image file

Fractal			JPEG			JPEG 2000			2-Way Hybrid		3-Way Hybrid	
MSE	CR	PSNR	Quality	CR	PSNR	Qs	CR	PSNR	CR	PSNR	CR	PSNR
6.53	9.202	36.543	68	9.208	36.496	0.3	9.192	67.606	9.368	39.711	9.207	67.606
7.54	10.484	35.836	61	10.478	35.631	0.5	10.500	55.503	10.691	38.657	10.378	55.511
9.20	12.698	33.982	48	12.702	34.464	0.9	12.655	51.151	12.953	37.233	12.526	51.160
9.88	13.670	33.587	43	13.670	34.036	1.1	13.632	49.717	13.937	36.742	13.481	49.761
12.00	17.070	31.426	30	17.018	32.606	1.9	17.081	45.482	17.427	35.257	16.663	45.602
13.63	19.994	30.613	23	19.997	31.560	2.7	19.965	42.758	20.578	34.282	19.393	42.964
14.86	22.549	30.084	19	22.534	30.831	3.5	22.517	40.661	23.214	33.549	21.713	41.043
16.01	25.073	29.423	16	25.082	30.229	4.4	25.075	38.780	25.720	32.907	23.990	39.503
17.86	29.696	28.598	12	29.717	29.312	6.3	29.723	35.861	30.466	31.873	28.156	37.133
19.05	32.741	28.258	10	32.794	28.718	7.7	32.701	34.296	33.688	31.218	31.073	35.804
22.76	42.970	27.152	6	42.963	27.159	12.4	42.985	30.085	44.083	29.370	40.469	32.441
23.86	46.499	26.790	5	46.490	26.390	14.2	46.446	28.954	47.735	28.751	44.239	31.585
25.49	51.163	26.424	4	51.083	25.288	16.6	51.180	27.999	52.664	28.029	48.512	30.374
26.86	55.696	25.969	3	55.719	24.226	18.0	55.660	27.025	57.579	27.299	53.535	29.812
27.63	58.518	36.543	2	58.754	23.832	19.4	58.511	27.151	60.685	27.002	56.597	29.456

Figure 7.27 Original Kameraman512.pgm image.

Figure 7.28 Kameraman512.pgm using fractal $(MSE = 19.05, CR = 32.741, PSNR = 28.258)$.

Figure 7.29 Kameraman512.pgm using JPEG $(Quality = 10, CR = 32.794, PSNR = 28.718)$.

Figure 7.30 Kameraman512.pgm using 2-way hybrid$\left(CR = 33.688, PSNR = 31.218 \right)$.

Figure 7.31 Kameraman512.pgm using JPEG2000 $(Qs = 7.7, CR = 32.701, PSNR = 34.296)$.

Figure 7.32 Kameraman512.pgm using 3-way hybrid $(CR = 31.073, PSNR = 35.804)$.

7.13 Results for Lena128.pgm Image

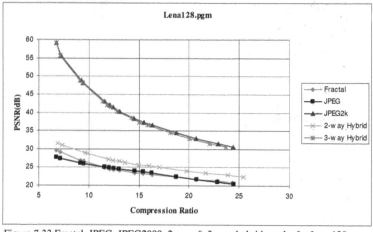

Figure 7.33 Fractal, JPEG, JPEG2000, 2-way & 3-way hybrid results for Lena128.pgm.

The graph shown in Figure 7.33 shows negligible improvement in CR when using the 3-way hybrid algorithm versus the JPEG2000 compression algorithm alone. However, a great deal of improvement can be realized when using the 3-way hybrid algorithm over the 2-way hybrid algorithm or over the fractal or JPEG algorithm alone. The image itself is an 8-bit grayscale image measuring 128x128 pixels as shown in part (a) of Figure 7.34. Using just the fractal algorithm, the compression ratio ranges from 6.757 to 24.367, the PSNR from 29.597 down to 20.923, with the MSE being varied from 14.35 to 32.83. If just the JPEG compression algorithm is used, the compression ratio varies from 6.757 to 24.440 and the PSNR from 27.673 to 20.459 for quality levels of 37 down to 2. If just the JPEG2000 compression algorithm is used, the compression ratio varies from 6.755 to 24.426 and the PSNR from 59.183 to 30.576 for explicit quantization step sizes of 0.4 up to 11.1. Compressed using the 2-way hybrid algorithm, the compression ratio varies from 6.919 to 25.464 and the PSNR from 31.581 to 22.458. Compressed using the 3-way hybrid algorithm, the compression ratio varies from 6.749 to 23.698 and the PSNR from 59.183 to 30.718. Overall, the average gain in PSNR using the 3-way hybrid algorithm over the 2-way hybrid algorithm is approximately 55.2%. The chart shown in Figure 7.33 was generated from the data listed in Table 7.13.

Figure 7.34 shows examples of the Lena128.pgm image compressed by each method. For this comparison, the image samples were all selected at the data points in Figure 7.33 where CR is approximately 14:1. Part (a) of Figure 7.34 is the original image. Part (b) of Figure 7.34 is the image compressed by fractal with MSE=26.37, CR=14.551, and PSNR=23.485. Part (c) of Figure 7.34 is the image compressed by JPEG with Quality=9, CR=14.59, and PSNR=24.054. Part (d) of Figure 7.34 is the image compressed by the 2-way hybrid with CR=14.963 and PSNR=25.647. Part (e) of Figure 7.34 is the image compressed by JPEG2000 with Qs=4.2, CR=14.534, and PSNR=38.421. Part (f) of Figure 7.34 is the image compressed by the 3-way hybrid with CR=14.248 and PSNR=38.433. Subjectively, the 3-way hybrid of part (f) is better than the other methods.

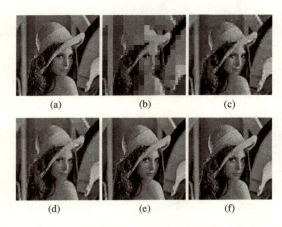

(a)	(b)	(c)
(d)	(e)	(f)

Figure 7.34 Lena128.pgm image at CR approximately 14:1.

Table 7.13 Compression algorithm results for Lena128.pgm image file

Fractal			JPEG			JPEG 2000			2-Way Hybrid		3-Way Hybrid	
MSE	CR	PSNR	Quality	CR	PSNR	Qs	CR	PSNR	CR	PSNR	CR	PSNR
14.35	6.757	29.597	37	6.757	27.673	0.4	6.755	59.183	6.919	31.581	6.749	59.183
15.25	7.149	29.062	33	7.149	27.319	0.5	7.194	55.590	7.331	31.085	7.121	55.590
19.28	9.151	26.867	21	9.151	26.088	1.2	9.215	48.765	9.607	28.946	9.111	48.765
19.91	9.463	26.673	20	9.468	25.960	1.3	9.439	48.155	9.879	28.716	9.387	48.155
23.33	11.589	24.918	14	11.581	25.093	2.4	11.579	43.130	11.961	27.163	11.349	43.130
23.84	12.085	24.475	13	12.032	24.928	2.7	12.087	42.062	12.442	26.867	11.806	42.062
24.26	12.471	24.346	12	12.471	24.731	2.9	12.426	41.491	13.088	26.605	12.238	41.491
24.86	13.057	24.157	11	13.025	24.529	3.3	13.099	40.347	13.587	26.343	12.802	40.347
26.37	14.551	23.485	9	14.590	24.054	4.2	14.534	38.421	14.963	25.647	14.248	38.433
27.04	15.442	23.239	8	15.413	23.793	4.8	15.479	37.274	15.983	25.342	15.017	37.298
27.95	16.317	23.052	7	16.285	23.443	5.3	16.339	36.545	17.065	24.992	15.891	36.562
29.33	18.720	22.417	5	18.742	22.407	6.9	18.767	34.412	19.782	24.043	18.161	34.480
30.71	20.758	21.761	4	20.732	21.759	8.3	20.700	32.900	21.692	23.537	20.048	32.984
31.97	22.840	21.415	3	22.840	21.067	9.9	22.883	31.477	23.905	22.982	22.071	31.587
32.83	24.367	20.923	2	24.440	20.459	11.1	24.426	30.576	25.464	22.458	23.698	30.718

7.14 Results for Lena256.pgm Image

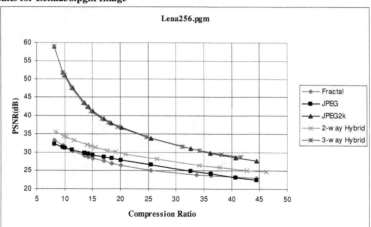

Figure 7.35 Fractal, JPEG, JPEG2000, 2-way & 3-way hybrid results for Lena256.pgm.

The graph shown in Figure 7.35 shows negligible improvement in CR when using the 3-way hybrid algorithm versus the JPEG2000 compression algorithm alone. However, a great deal of improvement can be realized when using the 3-way hybrid algorithm over the 2-way hybrid algorithm or over the fractal or JPEG algorithm alone. The image itself is an 8-bit grayscale image measuring 256x256 pixels as shown in part (a) of Figure 7.36. Using just the fractal algorithm, the compression ratio ranges from 8.174 to 44.502, the PSNR from 33.149 down to 23.011, with the MSE being varied from 9.13 to 27.97. If just the JPEG compression algorithm is used, the compression ratio varies from 8.182 to 44.472 and the PSNR from 32.229 to 22.397 for quality levels of 54 down to 1. If just the JPEG2000 compression algorithm is used, the compression ratio varies from 8.174 to 44.469 and the PSNR from 58.875 to 27.711 for explicit quantization step sizes of 0.4 up to 16.5. Compressed using the

2-way hybrid algorithm, the compression ratio varies from 8.343 to 46.293 and the PSNR from 35.395 to 24.628. Compressed using the 3-way hybrid algorithm, the compression ratio varies from 8.171 to 41.646 and the PSNR from 58.875 to 28.621. Overall, the average gain in PSNR using the 3-way hybrid algorithm over the 2-way hybrid algorithm is approximately 32.2%. The chart shown in Figure 7.35 was generated from the data listed in Table 7.14.

Figure 7.36 shows examples of the Lena256.pgm image compressed by each method. For this comparison, the image samples were all selected at the data points in Figure 7.35 where CR is approximately 20:1. Part (a) of Figure 7.36 is the original image. Part (b) of Figure 7.36 is the image compressed by fractal with MSE=19.09, CR=20.163, and PSNR=26.464. Part (c) of Figure 7.36 is the image compressed by JPEG with Quality=12, CR=20.139, and PSNR=27.811. Part (d) of Figure 7.36 is the image compressed by the 2-way hybrid with CR=20.943 and PSNR=29.46. Part (e) of Figure 7.36 is the image compressed by JPEG2000 with Qs=5.3, CR=20.176, and PSNR=36.658. Part (f) of Figure 7.36 is the image compressed by the 3-way hybrid with CR=19.503 and PSNR=38.816. Subjectively, the 3-way hybrid of part (f) is better than the other methods.

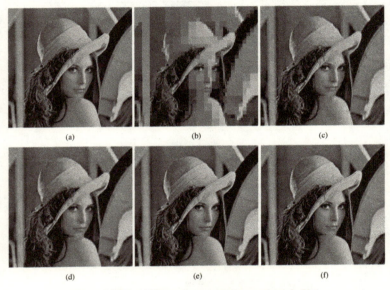

Figure 7.36 Lena256.pgm image at CR approximately 20:1.

Table 7.14 Compression algorithm results for Lena256.pgm image file

Fractal			JPEG			JPEG 2000			2-Way Hybrid		3-Way Hybrid	
MSE	CR	PSNR	Quality	CR	PSNR	Qs	CR	PSNR	CR	PSNR	CR	PSNR
9.13	8.174	33.149	54	8.182	32.229	0.4	8.174	58.875	8.343	35.395	8.171	58.875
10.77	9.697	31.755	41	9.703	31.326	0.8	9.690	51.694	9.935	34.310	9.633	51.694
11.03	10.005	31.367	39	10.005	31.155	0.9	10.007	51.017	10.262	34.081	9.927	51.017
12.48	11.392	30.174	32	11.327	30.522	1.4	11.397	47.553	11.668	33.144	11.179	47.553
14.60	13.555	28.963	24	13.560	29.727	2.3	13.562	43.497	13.989	32.011	13.275	43.497
15.19	14.241	28.618	22	14.235	29.490	2.6	14.228	42.437	14.701	31.688	13.995	42.437
15.81	15.118	28.297	20	15.128	29.218	3.0	15.104	41.246	15.600	31.319	14.774	41.250
17.32	17.106	27.617	16	17.111	28.665	3.9	17.086	39.132	17.631	30.500	16.739	39.173
18.22	18.424	27.038	14	18.429	28.289	4.5	18.412	37.980	19.145	30.029	17.861	38.062
19.09	20.163	26.464	12	20.139	27.811	5.3	20.176	36.658	20.943	29.460	19.503	36.816
21.64	25.476	25.127	8	25.476	26.571	7.7	25.535	33.761	26.517	28.135	24.718	34.116
24.85	33.685	23.863	5	32.677	24.925	11.0	32.690	30.905	34.141	26.507	31.111	31.407
25.58	36.236	23.650	4	36.236	24.139	12.7	36.228	29.736	37.825	25.871	34.212	30.474
26.74	40.715	23.370	3	40.791	23.044	14.7	40.737	28.555	42.732	25.082	38.222	29.380
27.97	44.502	23.011	1	44.472	22.397	16.5	44.469	27.711	46.293	24.628	41.646	28.621

7.15 Results for Lena512.pgm Image

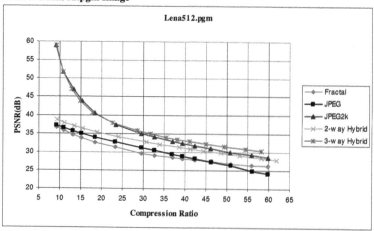

Figure 7.37 Fractal, JPEG, JPEG2000, 2-way & 3-way hybrid results for Lena512.pgm.

The graph shown in Figure 7.37 shows improvement in PSNR when using the 3-way hybrid algorithm versus the JPEG2000 compression algorithm alone at high CR. However, a great deal of improvement can be realized when using the 3-way hybrid algorithm over the 2-way hybrid algorithm or over the fractal or JPEG algorithm alone. The image itself is an 8-bit grayscale image measuring 512x512 pixels as shown in Figure 7.38. Using just the fractal algorithm, the compression ratio ranges from 9.182 to 59.867, the PSNR from 36.745 down to 26.244, with the MSE being varied from 5.67 to 19.84. If just the JPEG compression algorithm is used, the compression ratio varies from 9.189 to 59.895 and the PSNR from 37.204 to 24.242 for quality levels of 69 down to 2. If just the JPEG2000 compression algorithm is used, the compression ratio varies from 9.100 to 59.840 and the PSNR from 59.014 to 28.478 for explicit quantization step sizes of 0.4 up to 15.3. Compressed using the 2-way

hybrid algorithm, the compression ratio varies from 9.356 to 61.815 and the PSNR from 38.758 to 27.819. Compressed using the 3-way hybrid algorithm, the compression ratio varies from 9.151 to 58.296 and the PSNR from 59.014 to 30.255. Overall, the average gain in PSNR using the 3-way hybrid algorithm over the 2-way hybrid algorithm is approximately 16.1%. The chart shown in Figure 7.37 was generated from the data listed in Table 7.15.

For this comparison, the image samples were all selected at the data points in Figure 7.37 where CR is approximately 37:1. Figure 7.38 is the original image. Figure 7.39 is the image compressed by fractal with MSE=15.38, CR=36.976, and PSNR=28.495. Figure 7.40 is the image compressed by JPEG with Quality=8, CR=36.95, and PSNR=29.467. Figure 7.41 is the image compressed by the 2-way hybrid with CR=38.581 and PSNR=31.232. Figure 7.42 is the image compressed by JPEG2000 with Qs=8.8, CR=36.995, and PSNR=32.955. Figure 7.43 is the image compressed by the 3-way hybrid with CR=35.499 and PSNR=33.984. For this image, with the exception of the blockiness generated by fractal, none of the other methods showed excessive distortion. Subjectively, the 3-way hybrid of Figure 7.43 is better than the other methods.

Table 7.15 Compression algorithm results for Lena512.pgm image file

Fractal			JPEG			JPEG 2000			2-Way Hybrid		3-Way Hybrid	
MSE	CR	PSNR	Quality	CR	PSNR	Qs	CR	PSNR	CR	PSNR	CR	PSNR
5.67	9.182	36.745	69	9.189	37.204	0.4	9.100	59.014	9.356	38.758	9.151	59.014
6.41	10.864	35.813	60	10.887	36.454	0.8	10.830	51.730	11.089	37.975	10.708	51.730
7.37	13.159	34.661	46	13.153	35.586	1.5	13.207	46.989	13.421	37.093	12.857	46.989
8.21	15.230	33.680	37	15.224	34.901	2.2	15.284	43.861	15.526	36.357	14.830	43.854
9.54	18.478	32.508	27	18.448	33.951	3.3	18.498	40.542	18.889	35.344	17.842	40.610
11.26	23.334	31.241	18	23.326	32.619	4.9	23.420	37.442	24.060	34.058	22.619	37.837
13.36	29.509	29.613	12	29.549	31.095	6.7	29.477	35.043	30.752	32.635	28.651	35.763
14.27	32.733	29.090	10	32.733	30.411	7.6	32.638	34.078	33.976	32.015	31.677	34.928
15.38	36.976	28.495	8	36.950	29.467	8.8	36.995	32.955	38.581	31.232	35.499	33.984
15.91	39.446	28.153	7	39.375	28.894	9.5	39.447	32.387	41.207	30.784	38.199	33.469
16.56	42.373	27.887	6	42.386	28.240	10.3	42.348	31.708	44.449	30.342	41.084	32.910
17.53	46.228	27.402	5	46.277	27.328	11.4	46.152	30.890	48.539	29.689	45.029	32.233
18.46	50.865	26.858	4	50.954	26.467	12.7	50.705	29.997	53.155	29.067	49.399	31.506
19.30	56.005	26.458	3	56.065	24.824	14.2	55.945	29.163	58.348	28.290	54.458	30.743
19.84	59.867	26.244	2	59.895	24.242	15.3	59.840	28.478	61.815	27.819	58.296	30.255

Figure 7.38 Original Lena512.pgm image.

Figure 7.39 Lena512.pgm image using fractal $(MSE = 15.38, CR = 36.976, PSNR = 28.495)$.

Figure 7.40 Lena512.pgm image using JPEG $(Quality = 8, CR = 36.950, PSNR = 29.467)$.

Figure 7.41 Lena512.pgm image using 2-way hybrid $(CR = 38.581, PSNR = 31.232)$.

Figure 7.42 Lena512.pgm image using JPEG2000 $(Qs = 8.8, CR = 36.995, PSNR = 32.955)$.

Figure 7.43 Lena512.pgm image using 3-way hybrid $(CR = 35.499, PSNR = 33.984)$.

7.16 Results for Mosaic128.pgm Image

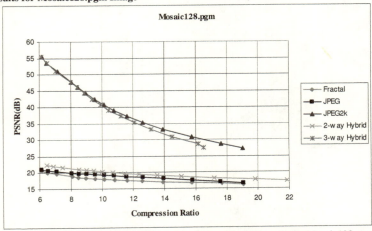

Figure 7.44 Fractal, JPEG, JPEG2000, 2-way & 3-way hybrid results for Mosaic128.pgm.

The graph shown in Figure 7.44 shows negligible improvement in CR when using the 3-way hybrid algorithm versus the JPEG2000 compression algorithm alone. However, a great deal of improvement can be realized when using the 3-way hybrid algorithm over the 2-way hybrid algorithm or over the fractal or JPEG algorithm alone. The image itself is an 8-bit grayscale image measuring 128x128 pixels as shown in part (a) of Figure 7.45. Using just the fractal algorithm, the compression ratio ranges from 6.142 to 19.069, the PSNR from 20.606 down to 16.239, with the MSE being varied from 33.23 to 51.32. If just the JPEG compression algorithm is used, the compression ratio varies from 6.142 to 19.069 and the PSNR from 21.003 to 16.621 for quality levels of 20 down to 1. If just the JPEG2000 compression algorithm is used, the compression ratio varies from 6.172 to 19.024 and the PSNR from 55.507 to 27.381 for explicit quantization step sizes of 0.5 up to 15.5. Compressed using the 2-way hybrid algorithm, the compression ratio varies from 6.451 to 21.865 and the PSNR from 22.520 to 17.415. Compressed using the 3-way hybrid algorithm, the compression ratio varies from 6.124 to 16.515 and the PSNR from 55.507 to 27.438. Overall, the average gain in PSNR using the 3-way hybrid algorithm over the 2-way hybrid algorithm is approximately 102.4%. The chart shown in Figure 7.44 was generated from the data listed in Table 7.16.

Figure 7.45 shows examples of the Mosaic128.pgm image compressed by each method. For this comparison, the image samples were all selected at the data points in Figure 7.44 where CR is approximately 13:1. Part (a) of Figure 7.45 is the original image. Part (b) of Figure 7.45 is the image compressed by fractal with MSE=46.71, CR=12.595, and PSNR=17.288. Part (c) of Figure 7.45 is the image compressed by JPEG with Quality=6, CR=12.605, and PSNR=18.443. Part (d) of Figure 7.45 is the image compressed by the 2-way hybrid with CR=13.519 and PSNR=19.201. Part (e) of Figure 7.45 is the image compressed by JPEG2000 with Qs=6, CR=12.622, and PSNR=35.373. Part (f) of Figure 7.45 is the image compressed by the 3-way hybrid with CR=12.085 and PSNR=35.376. Subjectively, the 3-way hybrid of part (f) is better than the other methods.

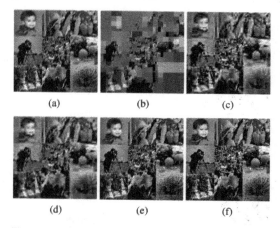

Figure 7.45 Mosaic128.pgm image at CR approximately 13:1.

Table 7.16 Compression algorithm results for Mosaic128.pgm image file

Fractal			JPEG			JPEG 2000			2-Way Hybrid		3-Way Hybrid	
MSE	CR	PSNR	Quality	CR	PSNR	Qs	CR	PSNR	CR	PSNR	CR	PSNR
33.23	6.142	20.606	20	6.142	21.003	0.5	6.172	55.507	6.451	22.520	6.124	55.507
34.56	6.523	20.109	18	6.528	20.728	0.6	6.473	53.637	6.896	22.120	6.513	53.637
36.44	7.102	19.654	16	7.102	20.414	0.9	7.176	50.993	7.478	21.679	7.065	50.993
38.84	8.074	18.921	13	8.070	19.952	1.4	8.043	47.664	8.528	21.102	7.992	47.664
39.64	8.492	18.611	12	8.506	19.791	1.7	8.466	46.074	8.996	20.862	8.427	46.074
40.83	8.981	18.410	11	8.996	19.611	2.1	8.974	44.261	9.490	20.635	8.903	44.261
41.91	9.579	18.168	10	9.579	19.431	2.6	9.542	42.477	10.018	20.337	9.392	42.477
43.20	10.160	17.968	9	10.148	19.261	3.2	10.154	40.754	10.704	20.060	9.957	40.754
44.36	10.796	17.783	8	10.789	19.014	3.9	10.790	39.020	11.444	19.792	10.452	39.020
45.53	11.606	17.495	7	11.589	18.753	4.8	11.608	37.261	12.386	19.517	11.248	37.261
46.71	12.595	17.288	6	12.605	18.443	6.0	12.622	35.373	13.519	19.201	12.085	35.376
47.84	13.921	17.053	5	13.945	18.047	7.7	13.902	33.188	15.086	18.719	13.161	33.198
49.49	15.753	16.738	4	15.783	17.577	10.3	15.742	30.789	17.172	18.225	14.512	30.806
50.67	17.671	16.505	3	17.614	17.033	13.2	17.636	28.726	19.806	17.769	16.189	28.753
51.32	19.069	16.239	1	19.069	16.621	15.5	19.024	27.381	21.865	17.415	16.515	27.438

7.17 Results for Mosaic256.pgm Image

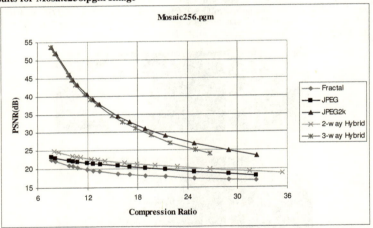

Figure 7.46 Fractal, JPEG, JPEG2000, 2-way & 3-way hybrid results for Mosaic256.pgm.

The graph shown in Figure 7.46 shows negligible improvement in CR when using the 3-way hybrid algorithm versus the JPEG2000 compression algorithm alone. However, a great deal of improvement can be realized when using the 3-way hybrid algorithm over the 2-way hybrid algorithm or over the fractal or JPEG algorithm alone. The image itself is an 8-bit grayscale image measuring 256x256 pixels as shown in part (a) of Figure 7.47. Using just the fractal algorithm, the compression ratio ranges from 7.672 to 32.243, the PSNR from 22.450 down to 16.932, with the MSE being varied from 28.28 to 48.68. If just the JPEG compression algorithm is used, the compression ratio varies from 7.632 to 32.259 and the PSNR from 23.252 to 18.170 for quality levels of 23 down to 2. If just the JPEG2000 compression algorithm is used, the compression ratio varies from 7.672 to 32.280 and the PSNR from 53.693 to 23.549 for explicit quantization step sizes of 0.6 up to 25.3. Compressed using the

140

2-way hybrid algorithm, the compression ratio varies from 7.930 to 35.395 and the PSNR from 24.898 to 18.885. Compressed using the 3-way hybrid algorithm, the compression ratio varies from 7.603 to 26.679 and the PSNR from 53.693 to 23.954. Overall, the average gain in PSNR using the 3-way hybrid algorithm over the 2-way hybrid algorithm is approximately 68%. The chart shown in Figure 7.46 was generated from the data listed in Table 7.17.

Figure 7.47 shows examples of the Mosaic256.pgm image compressed by each method. For this comparison, the image samples were all selected at the data points in Figure 7.46 where CR is approximately 17:1. Part (a) of Figure 7.47 is the original image. Part (b) of Figure 7.47 is the image compressed by fractal with MSE=40.79, CR=17.053, and PSNR=18.383. Part (c) of Figure 7.47 is the image compressed by JPEG with Quality=7, CR=17.044, and PSNR=20.542. Part (d) of Figure 7.47 is the image compressed by the 2-way hybrid with CR=17.94 and PSNR=21.305. Part (e) of Figure 7.47 is the image compressed by JPEG2000 with Qs=8.2, CR=17.069, and PSNR=32.756. Part (f) of Figure 7.47 is the image compressed by the 3-way hybrid with CR=16.177 and PSNR=32.851. Subjectively, the 3-way hybrid of part (f) is better than the other methods.

Figure 7.47 Mosaic256.pgm image at CR approximately 17:1.

Table 7.17 Compression algorithm results for Mosaic256.pgm image file

Fractal			JPEG			JPEG 2000			2-Way Hybrid		3-Way Hybrid	
MSE	CR	PSNR	Quality	CR	PSNR	Qs	CR	PSNR	CR	PSNR	CR	PSNR
28.28	7.672	22.450	23	7.632	23.252	0.6	7.672	53.693	7.930	24.898	7.603	53.693
29.19	8.126	22.037	21	8.125	23.003	0.8	8.178	51.802	8.486	24.540	8.075	51.802
32.33	9.775	20.880	16	9.776	22.326	1.7	9.821	46.048	10.183	23.552	9.670	46.048
33.21	10.234	20.655	15	10.234	22.175	2.0	10.265	44.680	10.673	23.332	10.075	44.682
34.03	10.746	20.415	14	10.744	22.010	2.4	10.799	43.165	11.175	23.112	10.561	43.169
35.69	11.921	19.848	12	11.921	21.661	3.3	11.953	40.445	12.413	22.637	11.672	40.447
36.72	12.664	19.561	11	12.625	21.484	3.9	12.669	39.050	13.181	22.386	12.319	39.060
37.66	13.457	19.370	10	13.491	21.295	4.6	13.457	37.673	14.034	22.130	13.089	37.702
39.51	15.652	18.716	8	15.652	20.839	6.7	15.639	34.476	16.433	21.617	14.922	34.536
40.79	17.053	18.383	7	17.044	20.542	8.2	17.069	32.756	17.940	21.305	16.177	32.851
42.08	18.907	18.183	6	18.902	20.207	10.2	18.927	30.967	20.089	20.953	17.712	31.056
43.67	21.331	17.938	5	21.352	19.767	13.0	21.344	29.042	22.737	20.489	19.603	29.136
45.64	24.802	17.449	4	24.811	19.223	17.1	24.796	26.782	26.658	19.912	21.916	26.912
47.25	28.966	17.120	3	28.979	18.579	21.8	28.950	24.816	31.485	19.287	24.991	25.060
48.68	32.243	16.932	2	32.259	18.170	25.3	32.280	23.549	35.395	18.885	26.679	23.954

7.18 Results for Mosaic512.pgm Image

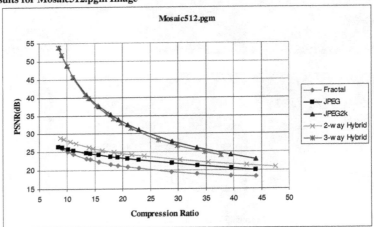

Figure 7.48 Fractal, JPEG, JPEG2000, 2-way & 3-way hybrid results for Mosaic512.pgm.

The graph shown in Figure 7.48 shows negligible improvement in CR when using the 3-way hybrid algorithm versus the JPEG2000 compression algorithm alone. However, a great deal of improvement can be realized when using the 3-way hybrid algorithm over the 2-way hybrid algorithm or over the fractal or JPEG algorithm alone. The image itself is an 8-bit grayscale image measuring 512x512 pixels as shown in Figure 7.49. Using just the fractal algorithm, the compression ratio ranges from 8.465 to 43.986, the PSNR from 26.403 down to 18.234, with the MSE being varied from 19.74 to 42.51. If just the JPEG compression algorithm is used, the compression ratio varies from 8.466 to 43.964 and the PSNR from 26.502 to 20.004 for quality levels of 30 down to 2. If just the JPEG2000 compression algorithm is used, the compression ratio varies from 8.501 to 43.995 and the PSNR from 53.706 to 22.953 for explicit quantization step sizes of 0.6 up to 27.9. Compressed using the 2-way hybrid algorithm, the compression ratio varies from 8.763 to 47.544 and the PSNR from 28.897 to 20.918. Compressed using the 3-way hybrid algorithm, the compression ratio varies from 8.433 to 37.775 and the PSNR from 53.706 to 24.029. Overall, the average gain in PSNR using the 3-way hybrid algorithm over the 2-way hybrid algorithm is approximately 46%. The chart shown in Figure 7.48 was generated from the data listed in Table 7.18.

For this comparison, the image samples were all selected at the data points in Figure 7.48 where CR is approximately 21:1. Figure 7.49 is the original image. Figure 7.50 is the image compressed by fractal with MSE=33.46, CR=20.869, and PSNR=20.863. Figure 7.51 is the image compressed by JPEG with Quality=8, CR=20.874, and PSNR=23.16. Figure 7.52 is the image compressed by the 2-way hybrid with CR=21.641 and PSNR=24.221. Figure 7.53 is the image compressed by JPEG2000 with Qs=8.6, CR=20.862, and PSNR=32.569. Figure 7.54 is the image compressed by the 3-way hybrid with CR=19.771 and PSNR=32.828. For this image, with the exception of the blockiness generated by fractal, none of the other methods showed excessive distortion. Subjectively, the 3-way hybrid of Figure 7.54 is better than the other methods.

Table 7.18 Compression algorithm results for Mosaic512.pgm image file

Fractal			JPEG			JPEG 2000			2-Way Hybrid		3-Way Hybrid	
MSE	CR	PSNR	Quality	CR	PSNR	Qs	CR	PSNR	CR	PSNR	CR	PSNR
19.74	8.465	26.403	30	8.466	26.502	0.6	8.501	53.706	8.763	28.897	8.433	53.706
20.90	9.074	25.953	27	9.077	26.176	0.8	9.055	51.761	9.396	28.449	9.033	51.762
22.62	10.124	25.125	23	10.133	25.714	1.2	10.000	48.850	10.480	27.798	10.044	48.851
24.19	11.183	24.416	20	11.184	25.333	1.8	11.152	45.574	11.541	27.210	11.036	45.588
27.08	13.506	23.136	15	13.511	24.650	3.2	13.510	40.736	13.951	26.220	13.212	40.770
27.78	14.137	22.928	14	14.142	24.484	3.6	14.126	39.760	14.607	25.980	13.811	39.796
29.32	15.725	22.283	12	15.725	24.123	4.7	15.747	37.528	16.276	25.475	15.195	37.606
31.14	17.809	21.612	10	17.825	23.696	6.2	17.802	35.254	18.437	24.914	17.117	35.421
32.30	19.188	21.172	9	19.183	23.440	7.3	19.199	33.891	19.876	24.579	18.268	34.085
33.46	20.869	20.863	8	20.874	23.160	8.6	20.862	32.569	21.641	24.221	19.771	32.828
34.72	22.924	20.482	7	22.904	22.830	10.3	22.937	31.078	23.835	23.810	21.480	31.387
37.75	28.844	19.532	5	28.859	21.938	15.5	28.859	27.784	30.339	22.801	26.419	28.202
39.49	33.439	18.952	4	33.456	21.267	19.5	33.424	26.031	35.451	22.138	29.865	26.536
41.37	39.535	18.440	3	39.589	20.458	24.3	39.521	24.097	42.318	21.381	34.783	24.874
42.51	43.986	18.234	2	43.964	20.004	27.9	43.995	22.953	47.544	20.918	37.775	24.029

143

Figure 7.49 Original Mosaic512.pgm image.

Figure 7.50 Mosaic512.pgm image using fractal $(MSE = 33.46, CR = 20.869, PSNR = 20.863)$.

Figure 7.51 Mosaic512.pgm image using JPEG $(Quality = 8, CR = 20.874, PSNR = 23.160)$.

Figure 7.52 Mosaic512.pgm image using 2-way hybrid $(CR = 21.641, PSNR = 24.221)$.

Figure 7.53 Mosaic512.pgm image using JPEG2000 $(Qs = 8.6, CR = 20.862, PSNR = 32.569)$.

Figure 7.54 Mosaic512.pgm image using 3-way hybrid $(CR = 19.771, PSNR = 32.828)$.

7.19 Results for Orchid128.pgm Image

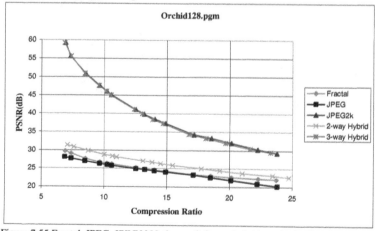

Figure 7.55 Fractal, JPEG, JPEG2000, 2-way & 3-way hybrid results for Orchid128.pgm.

The graph shown in Figure 7.55 shows negligible improvement in CR when using the 3-way hybrid algorithm versus the JPEG2000 compression algorithm alone. However, a great deal of improvement can be realized when using the 3-way hybrid algorithm over the 2-way hybrid algorithm or over the fractal or JPEG algorithm alone. The image itself is an 8-bit grayscale image measuring 128x128 pixels as shown in part (a) of Figure 7.56. Using just the fractal algorithm, the compression ratio ranges from 6.896 to 23.732, the PSNR from 29.703 down to 22.017, with the MSE being varied from 13.30 to 31.66. If just the JPEG compression algorithm is used, the compression ratio varies from 6.862 to 23.801 and the PSNR from 28.073 to 20.246 for quality levels of 34 down to 2. If just the JPEG2000 compression algorithm is used, the compression ratio varies from 6.896 to 23.788 and the PSNR from 59.230 to 29.323 for explicit quantization step sizes of 0.4 up to 13.1. Compressed using the 2-way hybrid algorithm, the compression ratio varies from 7.065 to 24.735 and the PSNR from 31.364 to 22.624. Compressed using the 3-way hybrid algorithm, the compression ratio varies from 6.856 to 23.261 and the PSNR from 59.230 to 29.530. Overall, the average gain in PSNR using the 3-way hybrid algorithm over the 2-way hybrid algorithm is approximately 52.6%. The chart shown in Figure 7.55 was generated from the data listed in Table 7.19.

Figure 7.56 shows examples of the Orchid128.pgm image compressed by each method. For this comparison, the image samples were all selected at the data points in Figure 7.55 where CR is approximately 17:1. Part (a) of Figure 7.56 is the original image. Part (b) of Figure 7.56 is the image compressed by fractal with MSE=26.92, CR=17.154, and PSNR=23.362. Part (c) of Figure 7.56 is the image compressed by JPEG with Quality=6, CR=17.118, and PSNR=23.222. Part (d) of Figure 7.56 is the image compressed by the 2-way hybrid with CR=17.748 and PSNR=25.023. Part (e) of Figure 7.56 is the image compressed by JPEG2000 with Qs=6.9, CR=17.165, and PSNR=34.249. Part (f) of Figure 7.56 is the image compressed by the 3-way hybrid with CR=16.751 and PSNR=34.368. Subjectively, the 3-way hybrid of part (f) is better than the other methods.

Figure 7.56 Orchid128.pgm image at CR approximately 17:1.

Table 7.19 Compression algorithm results for Orchid128.pgm image file

Fractal			JPEG			JPEG 2000			2-Way Hybrid		3-Way Hybrid	
MSE	CR	PSNR	Quality	CR	PSNR	Qs	CR	PSNR	CR	PSNR	CR	PSNR
13.30	6.896	29.703	34	6.862	28.073	0.4	6.896	59.230	7.065	31.364	6.856	59.230
14.12	7.360	29.061	30	7.364	27.704	0.5	7.290	55.577	7.575	30.886	7.337	55.577
16.40	8.497	27.635	23	8.457	26.948	0.9	8.501	50.888	8.732	29.805	8.440	50.891
18.52	9.641	26.673	18	9.641	26.277	1.4	9.625	47.655	10.005	28.820	9.584	47.659
19.03	10.237	26.351	16	10.262	25.948	1.7	10.211	46.047	10.663	28.362	10.179	46.091
19.51	10.614	26.168	15	10.614	25.764	1.9	10.569	45.114	10.991	28.119	10.539	44.987
22.30	12.528	25.155	11	12.547	24.905	3.1	12.514	41.101	13.088	27.073	12.395	41.151
23.23	13.236	24.808	10	13.246	24.652	3.6	13.200	39.788	13.827	26.723	13.036	39.937
24.18	14.016	24.583	9	14.016	24.346	4.2	14.018	38.505	14.512	26.332	13.862	38.349
25.15	14.935	24.261	8	14.922	24.021	4.9	14.918	37.268	15.471	25.931	14.694	37.291
26.92	17.154	23.362	6	17.118	23.222	6.9	17.165	34.249	17.748	25.023	16.751	34.368
27.88	18.488	23.022	5	18.488	22.576	8.1	18.560	33.303	19.225	24.448	18.100	33.339
29.31	20.146	22.621	4	20.121	21.838	9.5	20.159	32.021	21.024	23.755	19.663	32.060
30.88	22.281	22.300	3	22.221	20.875	11.5	22.246	30.331	23.427	23.106	21.635	30.455
31.66	23.732	22.017	2	23.801	20.246	13.1	23.788	29.323	24.735	22.624	23.261	29.530

7.20 Results for Orchid256.pgm Image

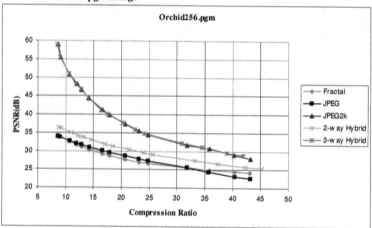

Figure 7.57 Fractal, JPEG, JPEG2000, 2-way & 3-way hybrid results for Orchid256.pgm.

The graph shown in Figure 7.57 shows negligible improvement in CR when using the 3-way hybrid algorithm versus the JPEG2000 compression algorithm alone. However, a great deal of improvement can be realized when using the 3-way hybrid algorithm over the 2-way hybrid algorithm or over the fractal or JPEG algorithm alone. The image itself is an 8-bit grayscale image measuring 256x256 pixels as shown in part (a) of Figure 7.58. Using just the fractal algorithm, the compression ratio ranges from 8.586 to 43.126, the PSNR from 34.335 down to 24.142, with the MSE being varied from 7.84 to 24.61. If just the JPEG compression algorithm is used, the compression ratio varies from 8.583 to 43.239 and the PSNR from 33.983 to 22.545 for quality levels of 51 down to 1. If just the JPEG2000 compression algorithm is used, the compression ratio varies from 8.595 to 43.244 and the PSNR from 59.043 to 28.076 for explicit quantization step sizes of 0.4 up to 16.2. Compressed using the

2-way hybrid algorithm, the compression ratio varies from 8.752 to 45.208 and the PSNR from 36.431 to 25.400. Compressed using the 3-way hybrid algorithm, the compression ratio varies from 8.563 to 41.859 and the PSNR from 59.016 to 28.955. Overall, the average gain in PSNR using the 3-way hybrid algorithm over the 2-way hybrid algorithm is approximately 30.7%. The chart shown in Figure 7.57 was generated from the data listed in Table 7.20.

Figure 7.58 shows examples of the Orchid256.pgm image compressed by each method. For this comparison, the image samples were all selected at the data points in Figure 7.57 where CR is approximately 32:1. Part (a) of Figure 7.58 is the original image. Part (b) of Figure 7.58 is the image compressed by fractal with MSE=21.09, CR=31.821, and PSNR=25.571. Part (c) of Figure 7.58 is the image compressed by JPEG with Quality=5, CR=31.836, and PSNR=25.488. Part (d) of Figure 7.58 is the image compressed by the 2-way hybrid with CR=33.275 and PSNR=27.393. Part (e) of Figure 7.58 is the image compressed by JPEG2000 with Qs=10.3, CR=31.86, and PSNR=31.691. Part (f) of Figure 7.58 is the image compressed by the 3-way hybrid with CR=30.862 and PSNR=32.159. Subjectively, the 3-way hybrid of part (f) is better than the other methods.

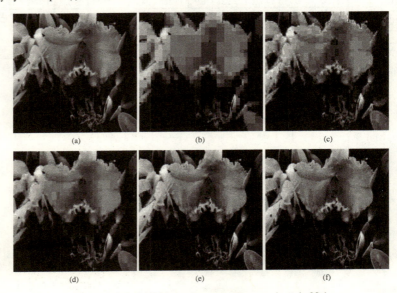

Figure 7.58 Orchid256.pgm image at CR approximately 32:1.

Table 7.20 Compression algorithm results for Orchid256.pgm image file

Fractal			JPEG			JPEG 2000			2-Way Hybrid		3-Way Hybrid	
MSE	CR	PSNR	Quality	CR	PSNR	Qs	CR	PSNR	CR	PSNR	CR	PSNR
7.84	8.586	34.335	51	8.583	33.983	0.4	8.595	59.043	8.752	36.431	8.563	59.016
8.27	9.028	33.925	46	9.028	33.692	0.5	9.080	55.417	9.190	36.105	8.985	55.441
9.68	10.687	32.565	34	10.695	32.662	0.9	10.676	50.828	10.975	34.891	10.624	50.911
10.71	11.968	31.578	28	11.953	32.014	1.3	12.001	48.207	12.218	34.139	11.830	48.172
11.42	12.883	31.088	25	12.886	31.613	1.6	12.879	46.608	13.176	33.653	12.711	46.737
12.44	14.219	30.316	21	14.225	31.023	2.1	14.191	44.267	14.622	32.963	14.049	44.469
13.98	16.612	29.242	16	16.604	30.077	3.1	16.621	41.261	17.088	31.888	16.388	41.268
14.75	17.915	28.638	14	17.930	29.587	3.7	17.938	39.771	18.522	31.356	17.669	39.932
16.08	20.672	27.651	11	20.646	28.662	5.0	20.682	37.415	21.408	30.396	20.213	37.656
17.31	23.187	26.886	9	23.220	27.841	6.2	23.154	35.463	24.038	29.595	22.495	35.809
18.17	24.755	26.567	8	24.755	27.356	7.0	24.756	34.455	25.676	29.108	24.153	34.838
21.09	31.821	25.571	5	31.836	25.488	10.3	31.860	31.691	33.275	27.393	30.862	32.159
22.58	35.762	24.945	4	35.723	24.408	12.1	35.802	30.767	37.287	26.559	34.501	31.104
23.97	40.364	24.484	3	40.389	23.158	14.8	40.317	28.969	42.236	25.816	38.926	29.639
24.61	43.126	24.142	1	43.239	22.545	16.2	43.244	28.076	45.208	25.400	41.859	28.955

7.21 Results for Orchid512.pgm Image

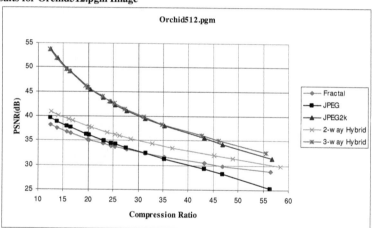

Figure 7.59 Fractal, JPEG, JPEG2000, 2-way & 3-way hybrid results for Orchid512.pgm.

The graph shown in Figure 7.59 shows negligible improvement in CR when using the 3-way hybrid algorithm versus the JPEG2000 compression algorithm alone. However, a great deal of improvement can be realized when using the 3-way hybrid algorithm over the 2-way hybrid algorithm or over the fractal or JPEG algorithm alone. The image itself is an 8-bit grayscale image measuring 512x512 pixels as shown in Figure 7.60. Using just the fractal algorithm, the compression ratio ranges from 12.443 to 56.500, the PSNR from 38.181 down to 28.771, with the MSE being varied from 5.19 to 14.49. If just the JPEG compression algorithm is used, the compression ratio varies from 12.439 to 56.306 and the PSNR from 39.634 to 25.238 for quality levels of 53 down to 3. If just the JPEG2000 compression algorithm is used, the compression ratio varies from 12.491 to 56.803 and the PSNR from 53.701 to 31.416 for explicit quantization step sizes of 0.6 up to 12.0. Compressed using the 2-way hybrid algorithm, the compression ratio varies from 12.646 to 58.452 and the PSNR from 40.911 to 29.809. Compressed using the 3-way hybrid algorithm, the compression ratio varies from 12.337 to 55.613 and the PSNR from 53.716 to 32.547. Overall, the average gain in PSNR using the 3-way hybrid algorithm over the 2-way hybrid algorithm is approximately 19.6%. The chart shown in Figure 7.59 was generated from the data listed in Table 7.21.

For this comparison, the image samples were all selected at the data points in Figure 7.59 where CR is approximately 43:1. Figure 7.60 is the original image. Figure 7.61 is the image compressed by fractal with MSE=12.28, CR=43.14, and PSNR=30.459. Figure 7.62 is the image compressed by JPEG with Quality=6, CR=43.225, and PSNR=29.328. Figure 7.63 is the image compressed by the 2-way hybrid with CR=45.122 and PSNR=31.992. Figure 7.64 is the image compressed by JPEG2000 with Qs=7.1, CR=43.141, and PSNR=35.542. Figure 7.65 is the image compressed by the 3-way hybrid with CR=42.711 and PSNR=36.075. For this image, with the exception of the blockiness generated by fractal, none of the other methods showed excessive distortion. Subjectively, the 3-way hybrid of Figure 7.65 is better than the other methods.

Table 7.21 Compression algorithm results for Orchid512.pgm image file

Fractal			JPEG			JPEG 2000			2-Way Hybrid		3-Way Hybrid	
MSE	CR	PSNR	Quality	CR	PSNR	Qs	CR	PSNR	CR	PSNR	CR	PSNR
5.19	12.443	38.181	53	12.439	39.634	0.6	12.491	53.701	12.646	40.911	12.337	53.716
5.55	13.834	37.490	45	13.814	38.864	0.8	13.856	51.827	14.082	40.177	13.682	51.808
6.07	15.759	36.674	36	15.743	37.930	1.1	15.774	49.590	16.097	39.329	15.582	49.585
6.25	16.455	36.434	34	16.477	37.627	1.2	16.407	49.075	16.848	39.029	16.313	49.089
7.13	19.877	35.210	25	19.646	36.281	1.8	19.900	45.901	20.141	37.779	19.426	46.167
7.19	20.143	35.097	24	20.101	36.093	1.9	20.433	45.356	20.621	37.651	19.904	45.762
7.89	22.986	34.334	19	23.104	34.958	2.4	22.984	43.837	23.744	36.668	22.862	44.002
8.32	24.528	33.928	17	24.533	34.419	2.7	24.503	42.935	25.327	36.189	24.229	43.046
8.55	25.396	33.739	16	25.450	34.218	2.9	25.390	42.251	26.300	35.965	25.195	42.611
9.06	27.668	33.215	14	27.642	33.512	3.4	27.696	41.099	28.468	35.347	27.283	41.400
9.92	31.419	32.427	11	31.494	32.370	4.2	31.406	39.426	32.782	34.430	31.109	39.834
10.70	35.227	31.635	9	35.232	31.292	5.1	35.247	38.028	36.872	33.542	34.677	38.279
12.28	43.140	30.459	6	43.225	29.328	7.1	43.141	35.542	45.122	31.992	42.711	36.075
12.98	46.831	29.826	5	46.822	28.299	8.2	46.799	34.245	48.929	31.343	46.033	35.053
14.49	56.500	28.771	3	56.306	25.238	12.0	56.803	31.416	58.452	29.809	55.613	32.547

Figure 7.60 Original Orchid512.pgm image.

Figure 7.61 Orchid512.pgm image using fractal $(MSE = 12.28, CR = 43.140, PSNR = 30.459)$.

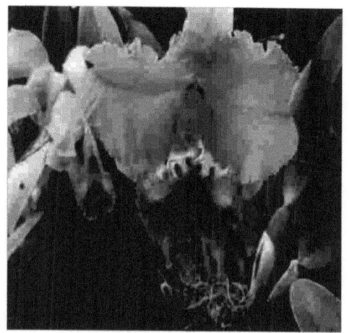

Figure 7.62 Orchid512.pgm image using JPEG $(Quality = 6, CR = 43.225, PSNR = 29.328)$.

Figure 7.63 Orchid512.pgm image using 2-way hybrid $(CR = 45.122, PSNR = 31.992)$.

Figure 7.64 Orchid512.pgm image using JPEG2000 $(Qs = 7.1, CR = 43.141, PSNR = 35.542)$.

Figure 7.65 Orchid512.pgm image using 3-way hybrid $(CR = 42.711, PSNR = 36.075)$.

7.22 Results for Peacock128.pgm Image

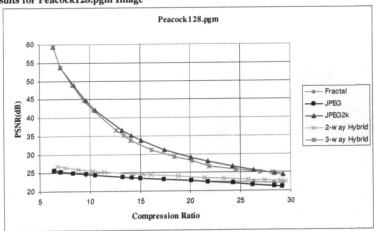

Figure 7.66 Fractal, JPEG, JPEG2000, 2-way & 3-way hybrid results for Peacock128.pgm.

The graph shown in Figure 7.66 shows negligible improvement in CR when using the 3-way hybrid algorithm versus the JPEG2000 compression algorithm alone. However, a great deal of improvement can be realized when using the 3-way hybrid algorithm over the 2-way hybrid algorithm or over the fractal or JPEG algorithm alone. The image itself is an 8-bit grayscale image measuring 128x128 pixels as shown in part (a) of Figure 7.67. Using just the fractal algorithm, the compression ratio ranges from 6.408 to 29.128, the PSNR from 25.839 down to 21.994, with the MSE being varied from 17.78 to 27.17. If just the JPEG compression algorithm is used, the compression ratio varies from 6.408 to 29.128 and the PSNR from 25.608 to 21.055 for quality levels of 37 down to 2. If just the JPEG2000 compression algorithm is used, the compression ratio varies from 6.398 to 29.192 and the PSNR from 59.358 to 24.397 for explicit quantization step sizes of 0.4 up to 21.7. Compressed using the 2-way hybrid algorithm, the compression ratio varies from 6.790 to 29.495 and the PSNR from 26.728 to 22.434. Compressed using the 3-way hybrid algorithm, the compression ratio varies from 6.383 to 28.177 and the PSNR from 59.358 to 24.804. Overall, the average gain in PSNR using the 3-way hybrid algorithm over the 2-way hybrid algorithm is approximately 47.6%. The chart shown in Figure 7.66 was generated from the data listed in Table 7.22.

Figure 7.67 shows examples of the Peacock128.pgm image compressed by each method. For this comparison, the image samples were all selected at the data points in Figure 7.66 where CR is approximately 20:1. Part (a) of Figure 7.67 is the original image. Part (b) of Figure 7.67 is the image compressed by fractal with MSE=25.08, CR=19.999, and PSNR=22.76. Part (c) of Figure 7.67 is the image compressed by JPEG with Quality=7, CR=20.072, and PSNR=22.757. Part (d) of Figure 7.67 is the image compressed by the 2-way hybrid with CR=21.409 and PSNR=23.55. Part (e) of Figure 7.67 is the image compressed by JPEG2000 with Qs=12.4, CR=20.011, and PSNR=29.13. Part (f) of Figure 7.67 is the image compressed by the 3-way hybrid with CR=18.426 and PSNR=29.188. Subjectively, the 3-way hybrid of part (f) is better than the other methods.

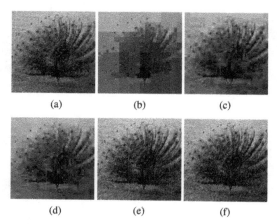

(a)   (b)   (c)

(d)   (e)   (f)

Figure 7.67 Peacock128.pgm image at CR approximately 20:1.

Table 7.22 Compression algorithm results for Peacock128.pgm image file

Fractal			JPEG			JPEG 2000			2-Way Hybrid		3-Way Hybrid	
MSE	CR	PSNR	Quality	CR	PSNR	Qs	CR	PSNR	CR	PSNR	CR	PSNR
17.78	6.408	25.839	37	6.408	25.608	0.4	6.398	59.358	6.790	26.728	6.383	59.358
18.59	7.108	25.464	32	7.108	25.303	0.6	7.056	53.747	7.529	26.384	7.075	53.747
19.75	8.371	24.965	26	8.350	24.949	1.2	8.378	48.785	8.845	25.911	8.241	48.785
20.74	9.618	24.541	21	9.618	24.615	2.0	9.608	44.752	10.275	25.480	9.430	44.752
21.39	10.539	24.307	18	10.546	24.371	2.7	10.524	42.032	11.420	25.193	10.307	42.032
22.85	13.268	23.804	13	13.279	23.804	5.2	13.213	36.535	14.322	24.616	12.615	36.535
23.12	14.161	23.581	12	14.174	23.660	6.1	14.106	35.082	15.255	24.441	13.387	35.082
23.57	15.059	23.419	11	15.017	23.534	7.1	15.083	33.820	16.062	24.284	14.064	33.821
24.50	17.390	23.122	9	17.390	23.193	9.7	17.416	31.127	18.806	23.928	16.141	31.161
25.08	19.999	22.760	7	20.072	22.757	12.4	20.011	29.130	21.409	23.550	18.426	29.188
25.51	21.807	22.495	6	21.749	22.453	14.1	21.701	27.989	23.327	23.276	20.196	28.117
26.05	24.295	22.356	5	24.259	22.131	16.7	24.201	26.544	25.744	22.998	21.836	26.682
26.37	26.280	22.255	4	26.238	21.597	18.8	26.288	25.611	27.423	22.741	24.549	25.821
26.82	28.323	22.125	3	28.372	21.233	21.0	28.358	24.674	28.973	22.574	26.928	25.040
27.17	29.128	21.994	2	29.128	21.055	21.7	29.192	24.397	29.495	22.434	28.177	24.804

7.23 Results for Peacock256.pgm Image

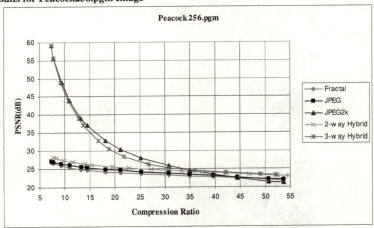

Figure 7.68 Fractal, JPEG, JPEG2000, 2-way & 3-way hybrid results for Peacock256.pgm.

The graph shown in Figure 7.68 shows improvement in CR when using the 3-way hybrid algorithm versus the JPEG2000 compression algorithm alone. However, a great deal of improvement can be realized when using the 3-way hybrid algorithm over the 2-way hybrid algorithm or over the fractal or JPEG algorithm alone. The image itself is an 8-bit grayscale image measuring 256x256 pixels as shown in part (a) of Figure 7.69. Using just the fractal algorithm, the compression ratio ranges from 7.314 to 53.730, the PSNR from 26.777 down to 22.255, with the MSE being varied from 16.04 to 26.48. If just the JPEG compression algorithm is used, the compression ratio varies from 7.312 to 53.642 and the PSNR from 27.079 to 21.829 for quality levels of 41 down to 1. If just the JPEG2000 compression algorithm is used, the compression ratio varies from 7.340 to 53.663 and the PSNR from 59.166 to 21.157 for explicit quantization step sizes of 0.4 up to 33.3. Compressed using the 2-way hybrid algorithm, the compression ratio varies from 7.769 to 54.626 and the PSNR from 28.135 to 22.971. Compressed using the 3-way hybrid algorithm, the compression ratio varies from 7.286 to 52.190 and the PSNR from 59.166 to 23.373. Overall, the average gain in PSNR using the 3-way hybrid algorithm over the 2-way hybrid algorithm is approximately 34.9%. The chart shown in Figure 7.68 was generated from the data listed in Table 7.23.

Figure 7.69 shows examples of the Peacock256.pgm image compressed by each method. For this comparison, the image samples were all selected at the data points in Figure 7.68 where CR is approximately 25:1. Part (a) of Figure 7.69 is the original image. Part (b) of Figure 7.69 is the image compressed by fractal with MSE=23.26, CR=25.27, and PSNR=23.7. Part (c) of Figure 7.69 is the image compressed by JPEG with Quality=9, CR=25.251, and PSNR=24.247. Part (d) of Figure 7.69 is the image compressed by the 2-way hybrid with CR=27.233 and PSNR=24.887. Part (e) of Figure 7.69 is the image compressed by JPEG2000 with Qs=14.1, CR=25.238, and PSNR=28.071. Part (f) of Figure 7.69 is the image compressed by the 3-way hybrid with CR=21.858 and PSNR=28.276. Subjectively, the 3-way hybrid of part (f) is better than the other methods.

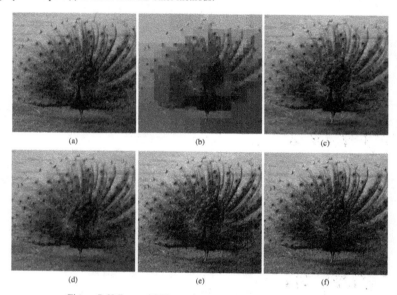

Figure 7.69 Peacock256.pgm image at CR approximately 25:1.

Table 7.23 Compression algorithm results for Peacock256.pgm image file

Fractal			JPEG			JPEG 2000			2-Way Hybrid		3-Way Hybrid	
MSE	CR	PSNR	Quality	CR	PSNR	Qs	CR	PSNR	CR	PSNR	CR	PSNR
16.04	7.314	26.777	41	7.312	27.079	0.4	7.340	59.166	7.769	28.135	7.286	59.166
16.34	7.705	26.578	38	7.728	26.912	0.5	7.693	55.577	8.198	27.936	7.680	55.577
17.59	9.358	25.906	30	9.343	26.437	1.2	9.370	48.854	9.803	27.337	9.207	48.854
18.54	11.011	25.462	24	11.013	26.034	2.2	11.020	43.869	11.647	26.878	10.762	43.869
19.69	13.267	24.953	19	13.299	25.614	3.9	13.264	38.986	14.073	26.374	12.775	38.986
20.17	14.486	24.735	17	14.496	25.425	4.9	14.485	37.021	15.402	26.160	13.768	37.023
21.38	18.143	24.251	13	18.118	24.946	8.1	18.144	32.663	19.451	25.615	16.731	32.712
22.23	21.118	23.984	11	21.173	24.628	10.8	21.156	30.262	22.792	25.272	18.793	30.391
23.26	25.270	23.700	9	25.251	24.247	14.1	25.238	28.071	27.233	24.887	21.858	28.276
24.15	30.804	23.230	7	30.833	23.782	18.6	30.833	25.853	33.410	24.427	26.745	26.280
24.81	34.905	22.949	6	34.923	23.459	21.7	34.925	24.588	37.436	24.119	30.717	25.304
25.39	39.632	22.716	5	39.656	23.084	25.0	39.605	23.554	42.182	23.769	35.977	24.522
25.73	44.562	22.513	4	44.472	22.546	28.2	44.431	22.474	46.789	23.408	42.074	23.946
26.27	50.619	22.313	3	50.658	22.023	31.9	50.675	21.449	52.525	23.080	49.028	23.509
26.48	53.730	22.255	1	53.642	21.829	33.3	53.663	21.157	54.626	22.971	52.190	23.373

7.24 Results for Peacock512.pgm Image

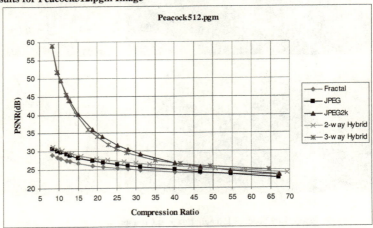

Figure 7.70 Fractal, JPEG, JPEG2000, 2-way & 3-way hybrid results for Peacock512.pgm.

The graph shown in Figure 7.70 shows negligible improvement in CR when using the 3-way hybrid algorithm versus the JPEG2000 compression algorithm alone. However, a great deal of improvement can be realized when using the 3-way hybrid algorithm over the 2-way hybrid algorithm or over the fractal or JPEG algorithm alone. The image itself is an 8-bit grayscale image measuring 512x512 pixels as shown in Figure 7.71. Using just the fractal algorithm, the compression ratio ranges from 8.233 to 67.169, the PSNR from 28.959 down to 23.573, with the MSE being varied from 12.29 to 23.31. If just the JPEG compression algorithm is used, the compression ratio varies from 8.205 to 67.186 and the PSNR from 30.775 to 22.815 for quality levels of 42 down to 1. If just the JPEG2000 compression algorithm is used, the compression ratio varies from 8.249 to 67.251 and the PSNR from 58.975 to 23.680 for explicit quantization step sizes of 0.4 up to 25.9. Compressed using the 2-way hybrid algorithm, the compression ratio varies from 8.405 to 69.263 and the PSNR from 31.268 to 24.237. Compressed using the 3-way hybrid algorithm, the compression ratio varies from 8.181 to 64.635 and the PSNR from 58.975 to 25.055. Overall, the average gain in PSNR using the 3-way hybrid algorithm over the 2-way hybrid algorithm is approximately 32%. The chart shown in Figure 7.70 was generated from the data listed in Table 7.24.

For this comparison, the image samples were all selected at the data points in Figure 7.70 where CR is approximately 28:1. Figure 7.71 is the original image. Figure 7.72 is the image compressed by fractal with MSE=19.13, CR=27.842, and PSNR=25.13. Figure 7.73 is the image compressed by JPEG with Quality=9, CR=27.901, and PSNR=26.131. Figure 7.74 is the image compressed by the 2-way hybrid with CR=29.937 and PSNR=26.743. Figure 7.75 is the image compressed by JPEG2000 with Qs=10.9, CR=27.868, and PSNR=30.464. Figure 7.76 is the image compressed by the 3-way hybrid with CR=24.913 and PSNR=30.624. For this image, with the exception of the blockiness generated by fractal, none of the other methods showed excessive distortion. Subjectively, the 3-way hybrid of Figure 7.76 is better than the other methods.

Table 7.24 Compression algorithm results for Peacock512.pgm image file

Fractal			JPEG			JPEG 2000			2-Way Hybrid		3-Way Hybrid	
MSE	CR	PSNR	Quality	CR	PSNR	Qs	CR	PSNR	CR	PSNR	CR	PSNR
12.29	8.233	28.959	42	8.205	30.775	0.4	8.249	58.975	8.405	31.268	8.181	58.975
13.21	9.567	28.304	33	9.566	30.074	0.8	9.598	51.718	9.813	30.558	9.481	51.718
13.76	10.486	27.964	29	10.493	29.665	1.1	10.403	49.496	10.805	30.146	10.355	49.496
14.56	12.006	27.444	24	12.027	29.097	1.8	11.955	45.497	12.462	29.592	11.748	45.497
14.91	12.806	27.232	22	12.810	28.824	2.2	12.795	43.821	13.305	29.324	12.477	43.821
15.80	15.015	26.735	18	15.023	28.210	3.4	15.084	40.144	15.703	28.734	14.482	40.146
17.04	18.740	26.122	14	18.714	27.445	5.6	18.763	35.936	19.707	27.984	17.632	35.958
17.75	21.382	25.747	12	21.383	26.991	7.1	21.306	33.947	22.606	27.546	19.879	33.993
18.60	25.217	25.337	10	25.195	26.450	9.4	25.216	31.676	26.833	27.034	22.818	31.783
19.13	27.842	25.130	9	27.901	26.131	10.9	27.868	30.464	29.937	26.743	24.913	30.624
19.64	31.102	24.907	8	31.113	25.793	12.7	31.057	29.228	33.580	26.404	27.558	29.469
20.94	40.184	24.364	6	40.178	24.941	17.2	40.146	26.811	44.023	25.684	34.959	27.378
21.63	46.789	24.110	5	46.814	24.415	19.7	46.774	25.783	50.571	25.235	41.448	26.536
22.37	54.651	23.870	4	54.651	23.733	22.5	54.622	24.742	58.413	24.848	49.278	25.814
23.31	67.169	23.573	1	67.186	22.815	25.9	67.251	23.680	69.263	24.237	64.635	25.055

Figure 7.71 Original Peacock512.pgm image.

Figure 7.72 Peacock512.pgm image using fractal $(MSE = 19.13, CR = 27.842, PSNR = 25.130)$.

Figure 7.73 Peacock512.pgm image using JPEG $(Quality = 9, CR = 27.901, PSNR = 26.131)$.

Figure 7.74 Peacock512.pgm image using 2-way hybrid $(CR = 29.337, PSNR = 26.743)$.

Figure 7.75 Peacock512.pgm image using JPEG2000 ($Qs = 10.9, CR = 27.868, PSNR = 30.464$).

Figure 7.76 Peacock512.pgm image using 3-way hybrid $(CR = 24.913, PSNR = 30.624)$.

7.25 Results for Peppers128.pgm Image

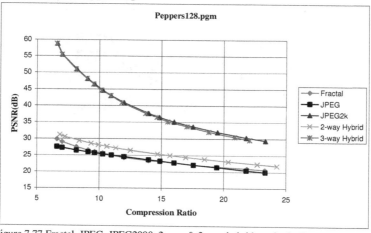

Figure 7.77 Fractal, JPEG, JPEG2000, 2-way & 3-way hybrid results for Peppers128.pgm.

The graph shown in Figure 7.77 shows negligible improvement in CR when using the 3-way hybrid algorithm versus the JPEG2000 compression algorithm alone. However, a great deal of improvement can be realized when using the 3-way hybrid algorithm over the 2-way hybrid algorithm or over the fractal or JPEG algorithm alone. The image itself is an 8-bit grayscale image measuring 128x128 pixels as shown in part (a) of Figure 7.78. Using just the fractal algorithm, the compression ratio ranges from 6.605 to 23.294, the PSNR from 29.864 down to 20.629, with the MSE being varied from 14.05 to 35.11. If just the JPEG compression algorithm is used, the compression ratio varies from 6.602 to 23.327 and the PSNR from 27.564 to 19.945 for quality levels of 33 down to 2. If just the JPEG2000 compression algorithm is used, the compression ratio varies from 6.635 to 23.339 and the PSNR from 58.792 to 29.473 for explicit quantization step sizes of 0.4 up to 12.6. Compressed using the 2-way hybrid algorithm, the compression ratio varies from 6.833 to 24.259 and the PSNR from 31.180 to 21.873. Compressed using the 3-way hybrid algorithm, the compression ratio varies from 6.589 to 22.071 and the PSNR from 58.792 to 29.682. Overall, the average gain in PSNR using the 3-way hybrid algorithm over the 2-way hybrid algorithm is approximately 55.9%. The chart shown in Figure 7.77 was generated from the data listed in Table 7.25.

Figure 7.78 shows examples of the Peppers128.pgm image compressed by each method. For this comparison, the image samples were all selected at the data points in Figure 7.77 where CR is approximately 14:1. Part (a) of Figure 7.78 is the original image. Part (b) of Figure 7.78 is the image compressed by fractal with MSE=27.43, CR=13.968, and PSNR=23.392. Part (c) of Figure 7.78 is the image compressed by JPEG with Quality=8, CR=13.968, and PSNR=23.651. Part (d) of Figure 7.78 is the image compressed by the 2-way hybrid with CR=14.694 and PSNR=25.438. Part (e) of Figure 7.78 is the image compressed by JPEG2000 with Qs=4.6, CR=13.971, and PSNR=37.607. Part (f) of Figure 7.78 is the image compressed by the 3-way hybrid with CR=13.723 and PSNR=37.614. Subjectively, the 3-way hybrid of part (f) is better than the other methods.

Figure 7.78 Peppers128.pgm image at CR approximately 14:1.

Table 7.25 Compression algorithm results for Peppers128.pgm image file

Fractal			JPEG			JPEG 2000			2-Way Hybrid		3-Way Hybrid	
MSE	CR	PSNR	Quality	CR	PSNR	Qs	CR	PSNR	CR	PSNR	CR	PSNR
14.05	6.605	29.864	33	6.602	27.564	0.4	6.635	58.792	6.833	31.180	6.589	58.792
15.06	7.026	29.030	30	7.029	27.255	0.5	7.018	55.406	7.237	30.604	6.990	55.406
18.11	8.175	27.544	23	8.171	26.505	0.9	8.186	51.006	8.405	29.413	8.114	51.006
20.20	9.105	26.520	19	9.121	25.991	1.3	9.062	48.129	9.355	28.594	9.005	48.129
21.30	9.629	26.030	17	9.629	25.694	1.6	9.635	46.469	9.927	28.165	9.540	46.469
22.50	10.282	25.631	15	10.288	25.349	2.0	10.283	44.603	10.614	27.679	10.179	44.603
23.63	10.933	24.996	13	10.977	25.018	2.4	10.943	43.019	11.325	27.204	10.911	43.019
24.95	11.988	24.281	11	11.988	24.568	3.1	11.985	40.864	12.499	26.509	11.806	40.864
27.43	13.968	23.392	8	13.968	23.651	4.6	13.971	37.607	14.694	25.438	13.723	37.614
28.09	14.895	23.125	7	14.895	23.255	5.3	14.847	36.441	15.678	24.977	14.642	36.450
29.52	16.062	22.729	6	16.015	22.754	6.2	16.008	35.084	16.941	24.427	15.485	35.109
30.55	17.558	22.045	5	17.558	22.086	7.4	17.523	33.680	18.447	23.885	17.029	33.703
32.21	19.476	21.562	4	19.453	21.390	9.1	19.505	31.979	20.371	23.161	18.785	32.042
34.20	21.865	21.071	3	21.807	20.477	11.1	21.802	30.387	22.682	22.392	20.864	30.568
35.11	23.294	20.629	2	23.327	19.945	12.6	23.339	29.473	24.259	21.873	22.071	29.682

7.26 Results for Peppers256.pgm Image

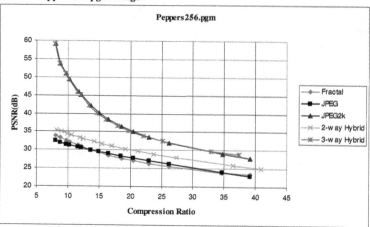

Figure 7.79 Fractal, JPEG, JPEG2000, 2-way & 3-way hybrid results for Peppers256.pgm.

The graph shown in Figure 7.79 shows negligible improvement in CR when using the 3-way hybrid algorithm versus the JPEG2000 compression algorithm alone. However, a great deal of improvement can be realized when using the 3-way hybrid algorithm over the 2-way hybrid algorithm or over the fractal or JPEG algorithm alone. The image itself is an 8-bit grayscale image measuring 256x256 pixels as shown in part (a) of Figure 7.80. Using just the fractal algorithm, the compression ratio ranges from 7.980 to 39.112, the PSNR from 33.882 down to 23.472, with the MSE being varied from 7.97 to 27.57. If just the JPEG compression algorithm is used, the compression ratio varies from 7.978 to 39.112 and the PSNR from 32.403 to 22.914 for quality levels of 52 down to 3. If just the JPEG2000 compression algorithm is used, the compression ratio varies from 7.985 to 39.138 and the PSNR from 59.110 to 27.867 for explicit quantization step sizes of 0.4 up to 15.9. Compressed using the 2-way

hybrid algorithm, the compression ratio varies from 8.147 to 40.893 and the PSNR from 35.487 to 25.076. Compressed using the 3-way hybrid algorithm, the compression ratio varies from 7.961 to 37.372 and the PSNR from 59.110 to 28.906. Overall, the average gain in PSNR using the 3-way hybrid algorithm over the 2-way hybrid algorithm is approximately 31.5%. The chart shown in Figure 7.79 was generated from the data listed in Table 7.26.

Figure 7.80 shows examples of the Peppers256.pgm image compressed by each method. For this comparison, the image samples were all selected at the data points in Figure 7.79 where CR is approximately 23:1. Part (a) of Figure 7.80 is the original image. Part (b) of Figure 7.80 is the image compressed by fractal with MSE=20.65, CR=22.944, and PSNR=26.266. Part (c) of Figure 7.80 is the image compressed by JPEG with Quality=9, CR=22.912, and PSNR=27.15. Part (d) of Figure 7.80 is the image compressed by the 2-way hybrid with CR=23.716 and PSNR=28.822. Part (e) of Figure 7.80 is the image compressed by JPEG2000 with Qs=7.7, CR=22.937, and PSNR=33.523. Part (f) of Figure 7.80 is the image compressed by the 3-way hybrid with CR=21.997 and PSNR=33.817. Subjectively, the 3-way hybrid of part (f) is better than the other methods.

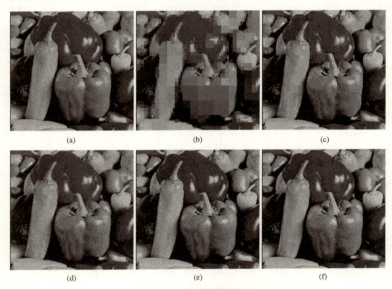

(a)　　　　　　　　　　(b)　　　　　　　　　　(c)

(d)　　　　　　　　　　(e)　　　　　　　　　　(f)

Figure 7.80 Peppers256.pgm image at CR approximately 23:1.

Table 7.26 Compression algorithm results for Peppers256.pgm image file

Fractal			JPEG			JPEG 2000			2-Way Hybrid		3-Way Hybrid	
MSE	CR	PSNR	Quality	CR	PSNR	Qs	CR	PSNR	CR	PSNR	CR	PSNR
7.97	7.980	33.882	52	7.978	32.403	0.4	7.985	59.110	8.147	35.487	7.961	59.110
8.70	8.758	33.294	45	8.758	31.950	0.6	8.760	53.688	8.922	34.981	8.695	53.688
9.62	9.693	32.553	38	9.693	31.469	0.9	9.682	50.966	9.949	34.397	9.605	50.966
10.19	10.250	32.137	35	10.254	31.235	1.1	10.228	49.474	10.522	34.078	10.160	49.474
11.58	11.602	31.201	29	11.602	30.690	1.7	11.634	45.961	11.882	33.292	11.462	45.961
12.10	12.081	30.913	27	12.088	30.495	1.9	12.063	45.061	12.408	33.027	11.973	45.061
13.52	13.679	29.895	22	13.676	29.903	2.7	13.695	42.080	14.155	32.247	13.405	42.081
14.77	14.990	29.307	19	14.939	29.479	3.4	14.995	40.145	15.431	31.663	14.583	40.151
16.10	16.449	28.468	16	16.441	28.968	4.2	16.469	38.398	16.987	31.067	16.031	38.421
17.79	18.538	27.681	13	18.512	28.342	5.3	18.555	36.513	19.206	30.283	17.999	36.576
19.12	20.408	27.040	11	20.434	27.811	6.3	20.405	35.124	21.173	29.628	19.750	35.266
20.65	22.944	26.266	9	22.912	27.150	7.7	22.937	33.523	23.716	28.822	21.997	33.817
22.38	26.158	25.565	7	26.210	26.224	9.4	26.173	32.004	27.370	27.920	25.193	32.454
26.10	34.701	23.866	4	34.701	23.992	13.7	34.799	29.016	36.478	25.911	32.924	29.770
27.57	39.112	23.472	3	39.112	22.914	15.9	39.138	27.867	40.893	25.076	37.372	28.906

7.27 Results for Peppers512.pgm Image

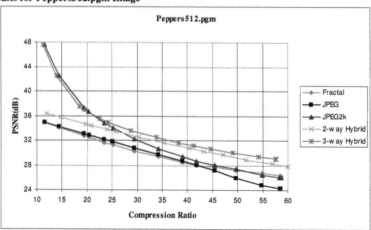

Figure 7.81 Fractal, JPEG, JPEG2000, 2-way & 3-way hybrid results for Peppers512.pgm.

The graph shown in Figure 7.81 shows a noticeable improvement in CR when using the 3-way hybrid algorithm versus the fractal, JPEG, JPEG2000, or 2-way hybrid compression algorithm alone. The image itself is an 8-bit grayscale image measuring 512x512 pixels as shown in Figure 7.82. Using just the fractal algorithm, the compression ratio ranges from 11.594 to 58.675, the PSNR from 34.909 down to 26.268, with the MSE being varied from 6.77 to 20.74. If just the JPEG compression algorithm is used, the compression ratio varies from 11.607 to 58.557 and the PSNR from 34.947 to 24.289 for quality levels of 54 down to 1. If just the JPEG2000 compression algorithm is used, the compression ratio varies from 11.610 to 58.698 and the PSNR from 47.589 to 26.032 for explicit quantization step sizes of 1.4 up to 20.5. Compressed using the 2-way hybrid algorithm, the compression ratio varies from 11.920 to 60.059 and the PSNR from 36.415 to 27.821. Compressed using the 3-way hybrid algorithm, the compression ratio varies from 11.337 to 57.655 and the PSNR from 47.589 to 29.033. Overall, the average gain in PSNR using the 3-way hybrid algorithm over the 2-way hybrid algorithm is approximately 7.1%. The chart shown in Figure 7.81 was generated from the data listed in Table 7.27.

For this comparison, the image samples were all selected at the data points in Figure 7.81 where CR is approximately 42:1. Figure 7.82 is the original image. Figure 7.83 is the image compressed by fractal with MSE=16.63, CR=41.832, and PSNR=28.082. Figure 7.84 is the image compressed by JPEG with Quality=6, CR=41.865, and PSNR=28.042. Figure 7.85 is the image compressed by the 2-way hybrid with CR=43.232 and PSNR=30.256. Figure 7.86 is the image compressed by JPEG2000 with Qs=14.6, CR=41.829, and PSNR=28.678. Figure 7.87 is the image compressed by the 3-way hybrid with CR=41.194 and PSNR=31.193. For this image, with the exception of the blockiness generated by fractal, none of the other methods showed excessive distortion. Subjectively, the 3-way hybrid of Figure 7.87 is better than the other methods.

Table 7.27 Compression algorithm results for Peppers512.pgm image file

Fractal			JPEG			JPEG 2000			2-Way Hybrid		3-Way Hybrid	
MSE	CR	PSNR	Quality	CR	PSNR	Qs	CR	PSNR	CR	PSNR	CR	PSNR
6.77	11.594	34.909	54	11.607	34.947	1.4	11.610	47.589	11.920	36.415	11.337	47.589
7.62	14.435	34.035	40	14.454	34.202	2.6	14.415	42.453	14.776	35.729	13.930	42.455
9.20	19.375	32.735	25	19.313	33.048	4.9	19.376	37.268	19.753	34.646	18.526	37.488
9.54	20.373	32.488	23	20.379	32.827	5.3	20.308	36.681	20.803	34.441	19.533	36.965
10.68	23.428	31.619	18	23.443	32.119	6.7	23.408	34.842	24.001	33.787	22.563	35.488
11.34	25.181	31.235	16	25.162	31.751	7.5	25.191	33.976	25.783	33.448	24.254	34.884
12.94	29.443	30.197	12	29.443	30.777	9.4	29.413	32.204	30.367	32.533	28.799	33.608
14.52	34.211	29.384	9	34.251	29.709	11.5	34.212	30.638	35.393	31.611	33.779	32.539
15.95	39.070	28.456	7	39.029	28.681	13.5	39.082	29.376	40.295	30.718	38.193	31.627
16.63	41.832	28.082	6	41.865	28.042	14.6	41.829	28.678	43.232	30.256	41.194	31.193
17.53	45.443	27.754	5	45.395	27.178	15.9	45.452	28.051	47.058	29.699	44.676	30.700
18.58	49.755	27.253	4	49.755	26.025	17.4	49.681	27.405	51.677	28.884	49.093	30.027
19.82	54.937	26.810	3	54.856	24.822	19.4	54.980	26.453	56.867	28.222	54.109	29.389
20.61	58.439	26.370	2	58.492	24.293	20.4	58.351	26.046	60.046	27.859	57.579	29.066
20.74	58.675	26.268	1	58.557	24.289	20.5	58.698	26.032	60.059	27.821	57.655	29.033

Figure 7.82 Original Peppers512.pgm image.

Figure 7.83 Peppers512.pgm image using fractal $(MSE = 16.63, CR = 41.832, PSNR = 28.082)$.

Figure 7.84 Peppers512.pgm image using JPEG $(Quality = 6, CR = 41.865, PSNR = 28.042)$.

Figure 7.85 Peppers512.pgm image using 2-way hybrid $\left(CR = 43.232, PSNR = 30.256\right)$.

Figure 7.86 Peppers512.pgm image using JPEG2000 $(Qs = 14.6, CR = 41.829, PSNR = 28.678)$.

Figure 7.87 Peppers512.pgm image using 3-way hybrid$(CR = 41.194, PSNR = 31.193)$.

7.28 Results for Waterfall128.pgm Image

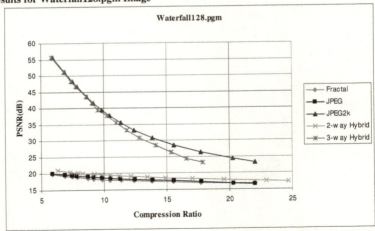

Figure 7.88 Fractal, JPEG, JPEG2000, 2-way & 3-way hybrid results for Waterfall128.pgm.

The graph shown in Figure 7.88 shows negligible improvement in CR when using the 3-way hybrid algorithm versus the JPEG2000 compression algorithm alone. However, a great deal of improvement can be realized when using the 3-way hybrid algorithm over the 2-way hybrid algorithm or over the fractal or JPEG algorithm alone. The image itself is an 8-bit grayscale image measuring 128x128 pixels as shown in part (a) of Figure 7.89. Using just the fractal algorithm, the compression ratio ranges from 5.903 to 22.012, the PSNR from 19.877 down to 16.705, with the MSE being varied from 34.09 to 47.88. If just the JPEG compression algorithm is used, the compression ratio varies from 5.869 to 22.042 and the PSNR from 20.180 to 16.517 for quality levels of 21 down to 2. If just the JPEG2000 compression algorithm is used, the compression ratio varies from 5.937 to 22.036 and the PSNR from 55.654 to 23.268 for explicit quantization step sizes of 0.5 up to 24.9. Compressed using the 2-way hybrid algorithm, the compression ratio varies from 6.302 to 24.697 and the PSNR from 21.058 to 17.337. Compressed using the 3-way hybrid algorithm, the compression ratio varies from 5.848 to 17.864 and the PSNR from 55.654 to 23.289. Overall, the average gain in PSNR using the 3-way hybrid algorithm over the 2-way hybrid algorithm is approximately 94.4%. The chart shown in Figure 7.88 was generated from the data listed in Table 7.28.

Figure 7.89 shows examples of the Peppers128.pgm image compressed by each method. For this comparison, the image samples were all selected at the data points in Figure 7.88 where CR is approximately 10:1. Part (a) of Figure 7.89 is the original image. Part (b) of Figure 7.89 is the image compressed by fractal with MSE=40.58, CR=9.832, and PSNR=18.206. Part (c) of Figure 7.89 is the image compressed by JPEG with Quality=10, CR=9.826, and PSNR=18.767. Part (d) of Figure 7.89 is the image compressed by the 2-way hybrid with CR=10.472 and PSNR=19.463. Part (e) of Figure 7.89 is the image compressed by JPEG2000 with Qs=3.7, CR=9.824, and PSNR=39.518. Part (f) of Figure 7.89 is the image compressed by the 3-way hybrid with CR=9.534 and PSNR=39.518. Subjectively, the 3-way hybrid of part (f) is better than the other methods.

Figure 7.89 Waterfall128.pgm image at CR approximately 10:1.

Table 7.28 Compression algorithm results for Waterfall128.pgm image file

Fractal			JPEG			JPEG 2000			2-Way Hybrid		3-Way Hybrid	
MSE	CR	PSNR	Quality	CR	PSNR	Qs	CR	PSNR	CR	PSNR	CR	PSNR
34.09	5.903	19.877	21	5.869	20.180	0.5	5.937	55.654	6.302	21.058	5.848	55.654
36.47	6.879	19.345	17	6.879	19.685	0.9	6.855	51.130	7.314	20.490	6.847	51.130
37.31	7.434	19.027	15	7.434	19.445	1.3	7.494	48.331	7.922	20.251	7.380	48.331
38.20	7.884	18.841	14	7.813	19.321	1.6	7.885	46.645	8.278	20.072	7.746	46.645
39.24	8.672	18.559	12	8.695	19.067	2.3	8.635	43.590	9.213	19.816	8.604	43.590
39.92	9.187	18.418	11	9.146	18.922	2.9	9.188	41.587	9.814	19.653	9.055	41.587
40.58	9.832	18.206	10	9.826	18.767	3.7	9.824	39.518	10.472	19.463	9.534	39.518
41.41	10.492	18.014	9	10.505	18.608	4.6	10.481	37.621	11.209	19.310	10.262	37.621
42.29	11.349	17.860	8	11.325	18.418	5.9	11.348	35.476	12.138	19.136	10.977	35.476
43.20	12.414	17.668	7	12.433	18.193	7.7	12.410	33.211	13.442	18.871	11.832	33.211
44.20	13.827	17.490	6	13.851	17.948	10.2	13.850	30.826	14.908	18.588	12.792	30.826
45.27	15.544	17.267	5	15.588	17.666	13.3	15.593	28.505	17.029	18.289	14.101	28.505
46.35	17.729	17.008	4	17.748	17.301	17.4	17.722	26.255	19.546	17.919	15.369	26.255
47.23	20.271	16.840	3	20.296	16.837	21.9	20.252	24.295	22.936	17.579	16.548	24.300
47.88	22.012	16.705	2	22.042	16.517	24.9	22.036	23.268	24.697	17.337	17.864	23.289

7.29 Results for Waterfall256.pgm Image

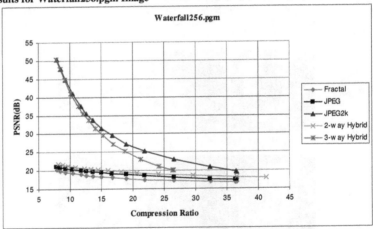

Figure 7.90 Fractal, JPEG, JPEG2000, 2-way & 3-way hybrid results for Waterfall256.pgm.

The graph shown in Figure 7.90 shows negligible improvement in CR when using the 3-way hybrid algorithm versus the fractal, JPEG, JPEG2000, or 2-way hybrid compression algorithm alone. The image itself is an 8-bit grayscale image measuring 256x256 pixels as shown in part (a) of Figure 7.91. Using just the fractal algorithm, the compression ratio ranges from 7.892 to 36.600, the PSNR from 20.127 down to 16.711, with the MSE being varied from 32.85 to 47.34. If just the JPEG compression algorithm is used, the compression ratio varies from 7.832 to 36.600 and the PSNR from 21.049 to 17.369 for quality levels of 19 down to 1. If just the JPEG2000 compression algorithm is used, the compression ratio varies from 7.895 to 36.612 and the PSNR from 50.308 to 19.564 for explicit quantization step sizes of 1.0 up to 39.9. Compressed using the 2-way hybrid algorithm, the compression ratio varies from 8.174 to 41.227 and the PSNR from 21.789 to 17.887. Compressed using the 3-way hybrid algorithm, the compression ratio varies from 7.774 to 26.432 and the PSNR from 50.308 to 19.968. Overall, the average gain in PSNR using the 3-way hybrid algorithm over the 2-way hybrid algorithm is approximately 61.5%. The chart shown in Figure 7.90 was generated from the data listed in Table 7.29.

Figure 7.91 shows examples of the Waterfall256.pgm image compressed by each method. For this comparison, the image samples were all selected at the data points in Figure 7.90 where CR is approximately 14:1. Part (a) of Figure 7.91 is the original image. Part (b) of Figure 7.91 is the image compressed by fractal with MSE=38.83, CR=13.622, and PSNR=18.524. Part (c) of Figure 7.91 is the image compressed by JPEG with Quality=9, CR=13.642, and PSNR=19.674. Part (d) of Figure 7.91 is the image compressed by the 2-way hybrid with CR=14.474 and PSNR=20.218. Part (e) of Figure 7.91 is the image compressed by JPEG2000 with Qs=7.3, CR=13.633, and PSNR=33.671. Part (f) of Figure 7.91 is the image compressed by the 3-way hybrid with CR=13.032 and PSNR=33.671. Subjectively, the 3-way hybrid of part (f) is better than the other methods.

Figure 7.91 Waterfall128.pgm image at CR approximately 14:1.

Table 7.29 Compression algorithm results for Waterfall256.pgm image file

Fractal			JPEG			JPEG 2000			2-Way Hybrid		3-Way Hybrid	
MSE	CR	PSNR	Quality	CR	PSNR	Qs	CR	PSNR	CR	PSNR	CR	PSNR
32.85	7.892	20.127	19	7.832	21.049	1.0	7.895	50.308	8.174	21.789	7.774	50.308
33.69	8.480	19.865	17	8.480	20.823	1.4	8.502	47.670	8.857	21.545	8.388	47.670
34.62	9.266	19.584	15	9.270	20.585	2.0	9.263	44.758	9.693	21.290	9.159	44.758
35.83	10.365	19.246	13	10.329	20.325	3.1	10.366	41.016	10.803	20.967	10.116	41.016
37.19	11.733	18.927	11	11.720	20.022	4.7	11.732	37.438	12.329	20.610	11.363	37.438
37.97	12.587	18.686	10	12.589	19.850	5.9	12.612	35.450	13.296	20.425	12.090	35.450
38.83	13.622	18.524	9	13.642	19.674	7.3	13.633	33.671	14.474	20.218	13.032	33.671
39.80	15.021	18.236	8	15.035	19.462	9.4	15.052	31.460	16.051	19.964	14.109	31.460
40.87	16.744	18.040	7	16.718	19.220	11.9	16.724	29.467	18.078	19.700	15.262	29.468
41.92	18.978	17.691	6	18.995	18.936	15.6	18.984	27.172	20.588	19.383	16.886	27.178
43.32	21.894	17.452	5	21.923	18.585	19.8	21.907	25.201	24.055	19.011	18.788	25.226
44.92	26.593	17.197	4	26.593	18.157	26.2	26.546	22.944	29.581	18.585	21.276	23.015
46.44	32.355	16.835	3	32.307	17.658	34.2	32.332	20.835	36.539	18.140	24.189	21.021
47.29	36.397	16.715	2	36.377	17.371	39.6	36.414	19.622	41.253	17.892	26.357	20.015
47.34	36.600	16.711	1	36.600	17.369	39.9	36.612	19.564	41.227	17.887	26.432	19.968

7.30 Results for Waterfall512.pgm Image

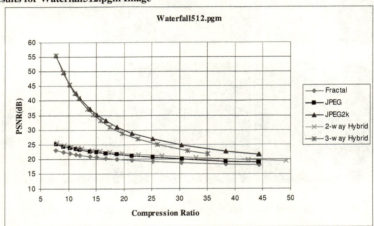

Figure 7.92 Fractal, JPEG, JPEG2000, 2-way & 3-way hybrid results for Waterfall512.pgm.

The graph shown in Figure 7.92 shows negligible improvement in CR when using the 3-way hybrid algorithm versus the fractal, JPEG, JPEG2000, or 2-way hybrid compression algorithm alone. The image itself is an 8-bit grayscale image measuring 512x512 pixels as shown in Figure 7.93. Using just the fractal algorithm, the compression ratio ranges from 7.746 to 44.351, the PSNR from 23.021 down to 17.964, with the MSE being varied from 23.43 to 41.13. If just the JPEG compression algorithm is used, the compression ratio varies from 7.762 to 44.366 and the PSNR from 25.070 to 18.767 for quality levels of 23 down to 1. If just the JPEG2000 compression algorithm is used, the compression ratio varies from 7.747 to 44.399 and the PSNR from 55.483 to 21.403 for explicit quantization step sizes of 0.5 up to 32.9. Compressed using the 2-way hybrid algorithm, the compression ratio varies from 7.921 to 49.232 and the PSNR from 25.480 to 19.373. Compressed using the 3-way hybrid algorithm, the compression ratio varies from 7.736 to 35.086 and the PSNR from 55.483 to 21.708. Overall, the average gain in PSNR using the 3-way hybrid algorithm over the 2-way hybrid algorithm is approximately 52.3%. The chart shown in Figure 7.92 was generated from the data listed in Table 7.30.

For this comparison, the image samples were all selected at the data points in Figure 7.92 where CR is approximately 17:1. Figure 7.93 is the original image. Figure 7.94 is the image compressed by fractal with MSE=32.14, CR=16.725, and PSNR=20.232. Figure 7.95 is the image compressed by JPEG with Quality=8, CR=16.73, and PSNR=21.973. Figure 7.96 is the image compressed by the 2-way hybrid with CR=17.439 and PSNR=22.358. Figure 7.97 is the image compressed by JPEG2000 with Qs=7.7, CR=16.695, and PSNR=33.154. Figure 7.98 is the image compressed by the 3-way hybrid with CR=15.853 and PSNR=33.155. For this image, with the exception of the blockiness generated by fractal, none of the other methods showed excessive distortion. Subjectively, the 3-way hybrid of Figure 7.98 is better than the other methods.

Table 7.30 Compression algorithm results for Waterfall512.pgm image file

Fractal			JPEG			JPEG 2000			2-Way Hybrid		3-Way Hybrid	
MSE	CR	PSNR	Quality	CR	PSNR	Qs	CR	PSNR	CR	PSNR	CR	PSNR
23.43	7.746	23.021	23	7.762	25.070	0.5	7.747	55.483	7.921	25.480	7.736	55.483
25.36	9.102	22.329	18	9.085	24.364	1.1	9.103	49.541	9.287	24.765	9.016	49.541
26.77	10.273	21.867	15	10.277	23.832	1.8	10.226	45.541	10.537	24.217	10.157	45.541
27.82	11.306	21.505	13	11.328	23.421	2.6	11.302	42.433	11.660	23.804	11.134	42.397
28.52	12.015	21.301	12	12.068	23.173	3.2	12.013	40.671	12.453	23.555	11.807	40.672
30.11	13.893	20.857	10	13.892	22.641	4.9	13.883	37.007	14.339	23.023	13.425	36.986
31.01	15.102	20.580	9	15.121	22.327	6.1	15.102	35.133	15.664	22.709	14.491	35.135
32.14	16.725	20.232	8	16.730	21.973	7.7	16.695	33.154	17.439	22.358	15.853	33.155
33.27	18.695	19.950	7	18.712	21.586	9.9	18.694	31.042	19.655	21.986	17.439	31.038
34.67	21.444	19.562	6	21.427	21.139	12.8	21.454	28.895	22.657	21.555	19.639	28.898
36.16	25.193	19.172	5	25.212	20.621	16.4	25.193	26.872	26.800	21.051	22.524	26.901
37.93	30.406	18.765	4	30.441	19.971	21.3	30.411	24.765	33.013	20.448	26.085	24.848
39.87	38.316	18.222	3	38.333	19.199	28.2	38.359	22.588	42.373	19.773	31.408	22.774
41.11	44.276	17.966	2	44.254	18.771	32.8	44.274	21.423	49.121	19.377	35.053	21.726
41.13	44.351	17.964	1	44.366	18.767	32.9	44.399	21.403	49.232	19.373	35.086	21.708

Figure 7.93 Original Waterfall512.pgm image.

Figure 7.94 Waterfall512.pgm image using fractal $(MSE = 32.14, CR = 16.725, PSNR = 20.232)$.

Figure 7.95 Waterfall512.pgm image using JPEG $(Quality = 8, CR = 16.730, PSNR = 21.973)$.

Figure 7.96 Waterfall512.pgm image using 2-way hybrid $(CR = 17.439, PSNR = 22.358)$.

Figure 7.97 Waterfall512.pgm image using JPEG2000 $(Qs = 7.7, CR = 16.695, PSNR = 33.154)$.

Figure 7.98 Waterfall512.pgm image using 3-way hybrid $(CR = 15.853, PSNR = 33.155)$.

7.31 Conclusions

The searchless fractal, DCT-based JPEG, and Wavelet-based JPEG 2000 image compression algorithms each provide a useful but different CR to PSNR relationship for the test images used here[30]. As demonstrated with the test images, the 3-way hybrid image compression method exceeds the capabilities of any of the component compression methods alone. Each algorithm is particularly well suited for certain image types and components. Building on the previously presented 2-way hybrid algorithm, the 3-way hybrid image compression algorithm combines the best capabilities of each of the three component algorithms.

The 3-way hybrid algorithm provides superior PSNR to CR performance when compared to the fractal, JPEG or JPEG 2000 algorithms individually for all tested CR. For those CR where the JPEG 2000 algorithm outperforms all of the other component methods by a PSNR of more than 5dB, the 3-way hybrid parallels the performance of JPEG 2000 with slightly lower CR due to the mosaic overhead associated with the hybrid. For CR where JPEG 2000 has similar or lower PSNR than the other components, the 3-way hybrid outperforms any of the component algorithms. This is especially notable at high CR where the 3-way hybrid exhibits the highest PSNR [30]. The 3-way hybrid also provides superior subjective image quality and collectively all of these factors make the 3-way hybrid image compression algorithm especially attractive for use in systems where high compression ratios are required.

REFERENCES

[1]   Rafael C. Gonzalez, Richard E. Woods, *Digital Image Processing*, 2nd Edition, Prentice-Hall, Inc., 2002.

[2]   A. N. Skodras, C. A. Christopoulos and T. Ebrahimi, *JPEG2000: The Upcoming Still Image Compression Standard.*

[3]   Viswanath Sankaranarayanan, Fractal Image Compression Project Report

[4]   Yuval Fisher, *Fractal Image Compression: Theory and Application.* Springer-Verlag New York, Inc., 1995.

[5]   J. E. Hutchinson, *Fractals and self-similarity*, Indiana University Mathematics Journal, 3(5): 713-747, 1981.

[6]   M. F. Barnsley, *Fractals Everywhere*, Academic Press, New York (1988).

[7]   Roger Stevens, *Fractal Programming in C*, M & T Publishing, Redwood City, California, 1989.

[8]   Michael Barnsley, Arnaud Jacquin, *"Application of Recurrent Iterated Function Systems to Images,"* Proceedings of the SPIE: Visual Communications and Image Processing, Vol. 1001, pp. 122-131, 1988.

[9]   Arnaud Jacquin, *"Image Coding Based on a Fractal Theory of Iterated Contracting Image Transformations,"* IEEE Transactions on Image Processing, Vol. 1, No. 1, pp. 18-30, January 1992.

[10]   E. Hamilton, *"JPEG File Interchange Format,"* http://www.jpeg.org/public/jfif.pdf, September 1992.

[11]   Rafael C. Gonzalez, Richard E. Woods, Steven L. Eddins, *Digital Image Processing using MATLAB*, Pearson Education, Inc. 2004.

[12]   Jason Elzinga, Keith Feenstra, *"JPEG 2000: The Next Compression Standard Using Wavelet Technology,"* http://www.gvsu.edu/math/wavelets/student_work/EF/index.html

[13]   Foley, van Dam, Feiner, Hughes, *Computer Graphics: Principles and Practice*, 2nd Edition in C, Addison-Wesley, 16th Printing, February 2002.

[14]   G. K. Wallace, *"The JPEG Still Picture Compression Standard,"* IEEE Transactions on Consumer Electronics, Vol. 38, pp. XVIII – XXXIV, February, 1992.

[15]   W. B. Pennebaker and J. L. Mitchell, *JPEG Still Image Data Compression Standard*, Van Nostrand Reinhold, 1993.

[16]   D. A. Huffman, *A Method for the construction of Minimum Redundancy Codes.* In Proceedings IRE, Vol. 40, 1962, pp. 1098-1101.

[17]   W. B. Pennebaker, J. L. Mitchell, et. Al. Arithmetic Coding articles. IBM J. Res. Dev., Vol. 32, No. 6 (Nov. 1988), pp. 717-774.

[18]   R. Hamzaoui, M. Muller, and D. Saupe, *"VQ-Enhanced Fractal Image Compression,"* Proceedings ICIP-96 (IEEE International Conference on Image Processing), Vol. I, pp. 153-156, Lausanne, Switzerland, September 1996.

[19]   N. T. Thao, *"A Hybrid Fractal-DCT Coding Scheme for Image Compression,"* Proceedings ICIP-96 (IEEE International Conference on Image Processing), Vol. I, pp. 169-172, Lausanne, Switzerland, September 1996.

[20]   G. Melnikov and A. K. Katsaggelos, *"A Non-Uniform Segmentation Optimal Hybrid Fractal-IDCT Image Compression Algorithm,"* Proceedings ICASSP-98 (IEEE International Conference on Acoustics, Speech and Signal Processing), Vol. 5, pp. 2573-2576, Seattle, WA, USA, May 1998.

[21]   K. M. Curtis, G. Neil, and V. Fotopoulos, *"A Hybrid Fractal/DCT Image Compression Method,"* Proceedings DSP-2002 (International Conference on Digital Signal Processing 14th), Vol. 2, pp. 1337-1340, 1-3 July 2002.

[22]   A. N. Skodras, *"Fast Discrete Cosine Transform Pruning,"* IEEE Transactions Signal Processing, Vol. 42, No. 7, pp. 1833-1837, July 1994.

[23]   W. A. Stapleton, M. A. McNees, K. G. Ricks, and J. Jackson, "An Improved Hybrid Fractal/DCT Image Compression Method," The 2005 International Conference of Imaging Science, Systems, and Technology, Las Vegas, NV,   June 26-29, 2005.

[24]   W. A. Stapleton, *"A Parallel Implementation of a Fractal Image Compression Algorithm Using The Parallel Virtual Machine (PVM) Environment,"* Dissertation Project 1997.

[25]   W. A. Stapleton, K. Sripaipan, D. J. Jackson, K. G. Ricks, "Hybrid Fractal/JPEG Image Compression: A Case Study," The 2004 International Conference on Imaging Science, Systems, and Technology, Las Vegas, NV, June 21-24, 2004.

[26]   W. A. Stapleton, D. J. Jackson, K. G. Ricks, X. Wu, K. Sripaipan, "A Novel Image Compression Algorithm Hybridizing a Searchless, Quadtree Recomposition Fractal Method With JPEG," The 2004 International Conference on Imaging Science, Systems, and Technology, Las Vegas, NV, June 21-24, 2004.

[27]   X. Wu, D. J. Jackson, H. C. Chen, W. A. Stapleton, K. G. Ricks, "A New Deeper Quadtree Searchless IFS Fractal Image Encoding Method," The 2004 International Conference on Imaging Science, Systems, and Technology, Las Vegas, NV, June 21-24, 2004.

[28]   Scott E. Umbaugh, *Computer vision and image processing: a practical approach using CVIPtools*, Prentice-Hall, Inc., 1998.

[29]   C. S. Tong, M. Pi, "Fast Fractal Image Encoding Based on Adaptive Search", IEEE Trans. Image Processing, Vol. 10, pp. 1269-1277, September 2001.

[30]   W. A. Stapleton, M. A. McNees, "Hybridizing Searchless Fractal Image Compression with JPEG and JPEG 2000", 2006.

[31]   W. A. Stapleton, M. A. McNees, "A Three-Component Hybrid Image Compression Method", The 2006 International Conference on Image Processing, Computer Vision, and Pattern Recognition, Las Vegas, NV, June 26-29, 2006.

APPENDIX A

```
%%%%%%%%%%%%%%%%%%%%%%%%%%%%%%%%%%%%%%%%%%%%%%%%%%%%%%%%%%%%%%%%%%%%%%%%%%%%%%%
% SIERPINSKITRIANGLE-Computes and displays Sierpinski's Triangle based on %
%   the number of iterations, n, entered by the user.                     %
%                                                                         %
% Input Parameters: none                                                  %
% Returned Parameters: none                                               %
%                                                                         %
% Written by:  Michael A. McNees                                          %
% Date:  05/21/05                                                         %
%%%%%%%%%%%%%%%%%%%%%%%%%%%%%%%%%%%%%%%%%%%%%%%%%%%%%%%%%%%%%%%%%%%%%%%%%%%%%%%
function SierpinskiTriangle
% Prompt the user for the number of iterations to perform. Default is 5
% iterations.
    iterations=inputdlg('# of iterations','Input n',1,{'5'});

% Round the user-entered value for the number of iterations to the nearest
% integer value greater than or equal to the user input.
    n=ceil(abs(str2num(char(iterations(1)))));

% Determine the present size of the screen.
    s=get(0,'ScreenSize');

% Define the location for the figure window to be displayed on the screen.
    set(gcf,'Position',[0 0 s(3) s(4)-70])

% Create the name of the figure window.
    set(gcf,'Name',strcat('Sierpinski Triangle:n =',num2str(n)))

% Call the sierpinski function to generate Sierpinski's Traingle.
    sierpinski(0,0,5^n,0,5^n/2,5^n*sin(pi/3),ceil(5^n/(2^n)));

function sierpinski(x1,y1,x2,y2,x3,y3,max)
% Define the equations for the three sides of the triangle.
    side_1 = sqrt( ( x1 - x2 ) ^2 + ( y1 - y2 ) ^2 );
    side_2 = sqrt( ( x1 - x3 ) ^2 + ( y1 - y3 ) ^2 );
    side_3 = sqrt( ( x3 - x2 ) ^2 + ( y3 - y2 ) ^2 );

    if( side_1 > max | side_2 > max | side_3 > max  )
        x6 = ( x1 + x2 ) / 2;
        y6 = ( y1 + y2 ) / 2;
        x5 = ( x2 + x3 ) / 2;
        y5 = ( y2 + y3 ) / 2;
        x4 = ( x1 + x3 ) / 2;
        y4 = ( y1 + y3 ) / 2;
        sierpinski(x1,y1,x4,y4,x6,y6,max)
        sierpinski(x2,y2,x6,y6,x5,y5,max)
        sierpinski(x3,y3,x4,y4,x5,y5,max)
    else
% Draw Sierpinski's Triangle on the screen.
        fill([x1,x2,x3],[y1,y2,y3],[0 0 1]);
        set(gca,'Visible','off')
        hold on;
    end
```

APPENDIX B

```
function y = im2jpeg2k(x, n, q)
%IM2JPEG2K Compresses an image using a JPEG 2000 approximation.
%   Y = IM2JPEG2K(X, N, Q) compresses image X using an N-scale JPEG
%   2K wavelet transform, implicit or explicit coefficient
%   quantization, and Huffman symbol coding augmented by zero
%   run-length coding. If quantization vector Q contains two
%   elements, they are assumed to be implicit quantization
%   parameters; else, it is assumed to contain explicit subband step
%   sizes. Y is an encoding structure containing Huffman-encoded
%   data and additional parameters needed by JPEG2K2IM for decoding.
%
%   See also JPEG2K2IM.

%   Copyright 2002-2004 R. C. Gonzalez, R. E. Woods, & S. L. Eddins
%   Digital Image Processing Using MATLAB, Prentice-Hall, 2004
%   $Revision: 1.5 $  $Date: 2003/10/26 18:38:13 $

global RUNS

%error(nargchk(3, 3, nargin));            % Check input arguments

if ndims(x) ~= 2 | ~isreal(x) | ~isnumeric(x) | ~isa(x, 'uint8')
   error('The input must be a UINT8 image.');
end

if length(q) ~= 2 & length(q) ~= 3 * n + 1
   error('The quantization step size vector is bad.');
end

% Level shift the input and compute its wavelet transform.
x = double(x) - 128;
[c, s] = wavefast(x, n, 'jpeg9.7');

% Quantize the wavelet coefficients.
q = stepsize(n, q);
sgn = sign(c);
sgn(find(sgn == 0)) = 1;
c = abs(c);

for k = 1:n
   qi = 3 * k - 2;
   c = wavepaste('h', c, s, k, wavecopy('h', c, s, k) / q(qi));
   c = wavepaste('v', c, s, k, wavecopy('v', c, s, k) / q(qi + 1));
   c = wavepaste('d', c, s, k, wavecopy('d', c, s, k) / q(qi + 2));
end
c = wavepaste('a', c, s, k, wavecopy('a', c, s, k) / q(qi + 3));
c = floor(c);
c = c .* sgn;

% Run-length code zero runs of more than 10. Begin by creating
% a special code for 0 runs ('zrc') and end-of-code ('eoc') and
% making a run-length table.
zrc = min(c(:)) - 1;
eoc = zrc - 1;
RUNS = [65535];

% Find the run transition points: 'plus' contains the index of the
% start of a zero run; the corresponding 'minus' is its end + 1.
z = c == 0;
z = z - [0 z(1:end - 1)];
plus = find(z == 1);
minus = find(z == -1);

% Remove any terminating zero run from 'c'.
if length(plus) ~= length(minus)
   c(plus(end):end) = [];
   c = [c eoc];
end

% Remove all other zero runs (based on 'plus' and 'minus') from 'c'.
```

197

```
for i = length(minus):-1:1
    run = minus(i) - plus(i);
    if run > 10
        ovrflo = floor(run / 65535);
        run = run - ovrflo * 65535;
        c = [c(1:plus(i) - 1) repmat([zrc 1], 1, ovrflo) zrc ...
             runcode(run) c(minus(i):end)];
    end
end

% Huffman encode and add misc. information for decoding.
y.runs    = uint16(RUNS);
y.s       = uint16(s(:));
y.zrc     = uint16(-zrc);
y.q       = uint16(100 * q');
y.n       = uint16(n);
y.huffman = mat2huff(c);
%---------------------------------------------------------------------%
function y = runcode(x)
% Find a zero run in the run-length table. If not found, create a
% new entry in the table. Return the index of the run.

global RUNS
y = find(RUNS == x);
if length(y) ~= 1
    RUNS = [RUNS; x];
    y = length(RUNS);
end
%---------------------------------------------------------------------%
function q = stepsize(n, p)
% Create a subband quantization array of step sizes ordered by
% decomposition (first to last) and subband (horizontal, vertical,
% diagonal, and for final decomposition the approximation subband).

if length(p) == 2                  % Implicit Quantization
    q = [];
    qn = 2 ^ (8 - p(2) + n) * (1 + p(1) / 2 ^ 11);
    for k = 1:n
        qk = 2 ^ -k * qn;
        q = [q (2 * qk) (2 * qk) (4 * qk)];
    end
    q = [q qk];
else                               % Explicit Quantization
    q = p;
end

q = round(q * 100) / 100;          % Round to 1/100th place
if any(100 * q > 65535)
    error('The quantizing steps are not UINT16 representable.');
end
if any(q == 0)
    error('A quantizing step of 0 is not allowed.');
end
```

APPENDIX C

```
function x = jpeg2k2im(y)
%JPEG2K2IM Decodes an IM2JPEG2K compressed image.
%    X = JPEG2K2IM(Y) decodes compressed image Y, reconstructing an
%    approximation of the original image X.  Y is an encoding
%    structure returned by IM2JPEG2K.
%
%    See also IM2JPEG2K.

%    Copyright 2002-2004 R. C. Gonzalez, R. E. Woods, & S. L. Eddins
%    Digital Image Processing Using MATLAB, Prentice-Hall, 2004
%    $Revision: 1.4 $  $Date: 2003/10/26 18:39:40 $

error(nargchk(1, 1, nargin));       % Check input arguments
% Get decoding parameters: scale, quantization vector, run-length
% table size, zero run code, end-of-data code, wavelet bookkeeping
% array, and run-length table.
n = double(y.n);
q = double(y.q) / 100;
runs = double(y.runs);
rlen = length(runs);
zrc = -double(y.zrc);
eoc = zrc - 1;
s = double(y.s);
s = reshape(s, n + 2, 2);

% Compute the size of the wavelet transform.
cl = prod(s(1, :));
for i = 2:n + 1
   cl = cl + 3 * prod(s(i, :));
end

% Perform Huffman decoding followed by zero run decoding.
r = huff2mat(y.huffman);
c = [];   zi = find(r == zrc);    i = 1;
for j = 1:length(zi)
   c = [c r(i:zi(j) - 1) zeros(1, runs(r(zi(j) + 1)))];
   i = zi(j) + 2;
end

zi = find(r == eoc);                % Undo terminating zero run
if length(zi) == 1                  % or last non-zero run.
   c = [c r(i:zi - 1)];
   c = [c zeros(1, cl - length(c))];
else
   c = [c r(i:end)];
end

% Denormalize the coefficients.
c = c + (c > 0) - (c < 0);
for k = 1:n
   qi = 3 * k - 2;
   c = wavepaste('h', c, s, k, wavecopy('h', c, s, k) * q(qi));
   c = wavepaste('v', c, s, k, wavecopy('v', c, s, k) * q(qi + 1));
   c = wavepaste('d', c, s, k, wavecopy('d', c, s, k) * q(qi + 2));
end
c = wavepaste('a', c, s, k, wavecopy('a', c, s, k) * q(qi + 3));

% Compute the inverse wavelet transform and level shift.
x = waveback(c, s, 'jpeg9.7', n);
x = uint8(x + 128);
```

199

APPENDIX D

```
function [varargout]=Hybrid(varargin);
% A searchless fractal/JPEG Hybrid image compression algorithm.
% Run with no parameters for details on the number and type of input
% arguments.

% Original code by Dr. William A. Stapleton
% Modified by Michael A. McNees

close all;
clear all;
fclose all;

warning off all
code_version=2.16;
close all

DEBUG=1;
MAKELOG=1;
log_file='Hybrid.log';

if nargin < 1
    disp('Usage: [varargout]=Hybrid(varargin)');
    disp(' ');
    disp('There can be up to 3 input arguments.');
    disp('1---the image to be compressed.');
    disp('2---the Mean Square Error(MSE) tolerance.');
    disp('3---the JPEG quality factor.');
    disp('If any parameters are omitted, the user will be prompted.');
end

% The following are general data definitions and structure assignments
global SIZE_BITS
global AVERAGE_BITS
global SCALE_BITS
global one_transform
global mse_tolerance

SIZE_BITS=3;    %number of bits to store size of range
AVERAGE_BITS=3; %minimum number of bits to use to store range average
SCALE_BITS=3;   %number of bits to store domain-to-range contrast scale

one_transform = struct('size',8,'average',0.0,'mse',0.0,'filled',0,...
    'scale',0.0);
temp_transform=one_transform;

%use open_image to select an image
%make sure selected image is correct form and compatible size

if nargin<1
    [raw_image original_file_size]=open_image;
    % Display the original file size in bytes...
    fprintf('Original FileSize: %d bytes.\n',original_file_size);
    figure('Name', 'Original Image');
    imshow(raw_image);
    if isempty(raw_image)
        disp('Cancel was selected...File Open Dialog Box...');
        return
    end
else
        if isgray(varargin{1})
        raw_image=varargin{1};
    elseif isrgb(varargin{1})
        if DEBUG ~=0
            disp('Warning, current algorithm is only defined for ');
            disp('grayscale images.');
            disp('Converting to grayscale');
        end
        raw_image=rgb2gray(varargin{1});
    elseif ischar(varargin{1})
        if DEBUG ~=0
```

```
                disp('Attempting to open ')
                disp(varargin{1})
                disp(' as image file');
            end
            raw_image=open_image(varargin{1});
        else
            disp('Do not know how to handle ')
            disp(varargin{1})
            disp(' as image');
            return
        end
    end

    if isrgb(raw_image)
        if DEBUG ~=0
            disp('Warning, current algorithm is only defined for grayscale');
            disp(' images.');
            disp('Converting to grayscale');
        end
        raw_image=rgb2gray(raw_image);
    end

    %Prompt the user for a fractal mean-square error tolerance.
    if nargin < 2
        mse_input=inputdlg('Enter the MSE tolerance level','MSE Tolerance',...
            1,{'4.0'});
        if isempty(mse_input)    % Check to see if the user clicked Cancel...
            disp('Cancel was selected...Input Dialog Box... MSE tolerance...');
            return
        end
        mse_tolerance=str2num(mse_input{1});
    else
        if isnumeric(varargin{2})
            mse_tolerance=varargin{2};
        elseif ischar(varargin{2})
            mse_tolerance=str2num(varargin{2});
        else
            disp('Do not know how to handle ');
            disp(varargin{2});
            disp(' as MSE tolerance');
            return
        end
    end

    %create a JPEG version of the image
    if nargin < 3
        jpeg_input=inputdlg('Enter the JPEG tolerance level',...
            'JPEG Tolerance',1,{'85'});
        if isempty(jpeg_input)    % Check to see if the user clicked Cancel...
            disp('Cancel was selected...JPEG tolerance input Dialog Box...');
            return
        end
        jpeg_tolerance=str2num(jpeg_input{1});
    else
        if isnumeric(varargin{3})
            jpeg_tolerance=varargin{3};
        elseif ischar(varargin{3})
            jpeg_tolerance=str2num(varargin{3});
        else
            disp('Do not know how to handle ');
            disp(varargin{3});
            disp(' as JPEG tolerance');
            return;
        end
    end

    tic; %start timer

    % Adjust the image size so that the compressed image is an integer
    % multiple of the largest range size.
```

```
[raw_height raw_width]=size(raw_image);
MAX_SIZE_BITS = log2(raw_height /4);
MAX_RANGE_SIZE=2^MAX_SIZE_BITS;

original_height=MAX_RANGE_SIZE*floor((raw_height+(MAX_RANGE_SIZE-1))/...
    MAX_RANGE_SIZE);
original_width=MAX_RANGE_SIZE*floor((raw_width+(MAX_RANGE_SIZE-1))/...
    MAX_RANGE_SIZE);
raw_average=sum(sum(raw_image))/(raw_height*raw_width);
original_image=raw_average*ones(original_height,original_width);
original_image(1:raw_height,1:raw_width)=raw_image;
original_image=double(original_image);

% output selected original image padded as necessary to a modulo-16 size
% create a structure array to store the fractal data.
% transforms=repmat(one_transform,original_height,original_width);
size_bits=0;
AVERAGE_QUANTIZER=(2^AVERAGE_BITS+size_bits);
fractal_size=ones([original_height original_width]);
fractal_average=AVERAGE_QUANTIZER*floor((original_image+...
    (AVERAGE_QUANTIZER/2))/AVERAGE_QUANTIZER);
fractal_mse=zeros([original_height original_width]);
fractal_scale=-1*ones([original_height original_width]);
fractal_filled=ones([original_height original_width]);

if DEBUG ~=0
    disp('creating 2x2, 4x4, 8x8, ... structures');
end

for size_bits=1:MAX_SIZE_BITS
    range_size=fix(2^size_bits);
    if DEBUG ~=0
        disp(range_size);
    end
    half_size=fix(range_size/2);
%     current_mse_tolerance=mse_tolerance*sqrt(range_size/2);
    for row=1:range_size:original_height
        for column=1:range_size:original_width
            if fractal_mse(row,column) <= mse_tolerance && ...
                    fractal_mse(row,column+half_size) <= mse_tolerance && ...
                    fractal_mse(row+half_size,column) <= mse_tolerance && ...
                    fractal_mse(row+half_size,column+half_size) <= mse_tolerance
                %create transform for row,column at range_size
                [temp_scale temp_mse temp_average]=...
                    create_transform(original_image,row,column,range_size);
                if temp_mse <= mse_tolerance
                    fractal_scale(row,column)=temp_scale;
                    fractal_mse(row,column)=temp_mse;
                    fractal_average(row,column)=temp_average;
                    fractal_size(row,column)=range_size;
                    fractal_filled(row,column)=1;
                    fractal_filled(row,column+half_size)=0;
                    fractal_filled(row+half_size,column)=0;
                    fractal_filled(row+half_size,column+half_size)=0;
                end
            end
        end
    end
end

% Generate an initial image for the fractal reconstruction from the stored
% average values for each range block.
fractal_image=128*ones([original_height,original_width]);
for row=1:original_height
    for column=1:original_width
        if fractal_filled(row,column) == 1
            fractal_image(row:(row+fractal_size(row,column)-1),...
                column:(column+fractal_size(row,column)-1))= ...
                fractal_average(row,column)*ones(fractal_size(row,column));
        end
    end
```

202

```
end

if DEBUG ~=0
    figure('Name','Fractal Image');
    imshow(uint8(fractal_image));
end

% Reconstruction of image from fractal data.
for iteration=1:5
    for row=1:original_height
        for column=1:original_width
            if fractal_filled(row,column) == 1
                if fractal_scale(row,column) == -1
                    new_image(row:row+fractal_size(row,column)-1,...
                        column:column+fractal_size(row,column)-1)=...
                        fractal_average(row,column)*...
                        ones(fractal_size(row,column));
                else
                    range_size=fractal_size(row,column);
                    domain_size=fix(2*range_size);
                    half_size=fix(range_size/2);
                    domain_row=row-half_size;
                    domain_column=column-half_size;
                    domain_row=fix(min(max(domain_row,1),...
                        (original_height-domain_size-1)));
                    domain_column=fix(min(max(domain_column,1),...
                        (original_width-domain_size-1)));
                    domain=(fractal_image(domain_row:2:...
                        (domain_row+domain_size-1),...
                        domain_column:2:(domain_column+domain_size-1))+...
                        fractal_image(domain_row+1:2:...
                        (domain_row+domain_size-1),domain_column:2:...
                        (domain_column+domain_size-1)) +...
                        fractal_image(domain_row:2:...
                        (domain_row+domain_size-1),...
                        domain_column+1:2:(domain_column+domain_size-1))+...
                        fractal_image(domain_row+1:2:...
                        (domain_row+domain_size-1),...
                        domain_column+1:2:(domain_column+domain_size-1)) )/4;
                    k= 0.0625+ (fractal_scale(row,column)/8.0);
                    new_image(row:row+range_size-1,...
                        column:column+range_size-1)=...
                        k*(domain-mean(mean(domain)))+...
                        fractal_average(row,column);
                end
            end
        end
    end
    fractal_image=new_image;
end

fractal_bits=0;
% Calculate the size of the fractal image file.
for row=1:original_height
    for column=1:original_width
        if fractal_filled(row,column)==1
            if fractal_size(row,column)==1
                fractal_bits=fractal_bits+5;
            else
                fractal_bits=fractal_bits+9;
            end
        end
    end
end
fractal_bytes=ceil(fractal_bits/8)+15;

fractal_compression_ratio = original_file_size / fractal_bytes;
fractal_rms=findrms(double(original_image),double(fractal_image));
fractal_psnr=20*log10(255/fractal_rms);

imwrite(uint8(original_image),'Hybrid_temp.jpg','jpeg',...
```

```
        'Quality',jpeg_tolerance);
jpeg_image=open_image('Hybrid_temp.jpg');
fid=fopen('Hybrid_temp.jpg','r');
    if fid == -1
        fprintf(2,'unable to open file Hybrid_temp.jpg\n');
        return
    end
jpeg_file=fread(fid);
fclose(fid);
jpeg_bytes=length(jpeg_file);
jpeg_compression_ratio = original_file_size / jpeg_bytes;
jpeg_file=[];

if DEBUG ~=0
    figure('Name','JPEG Image');
    imshow(jpeg_image);
end
jpeg_rms=findrms(double(original_image),double(jpeg_image));
jpeg_psnr=20*log10(255/jpeg_rms);

choosetype=zeros([original_height original_width]/8);
for row=1:8:original_height
    for column=1:8:original_width
        fractal_mse=findmse(original_image(row:row+7,column:column+7),...
            fractal_image(row:row+7,column:column+7));
        jpeg_mse=findmse(original_image(row:row+7,column:column+7),...
            double(jpeg_image(row:row+7,column:column+7)));
        if fractal_mse < jpeg_mse
            choosetype(1+fix(row/8),1+fix(column/8))=0;
        else
            choosetype(1+fix(row/8),1+fix(column/8))=1;
        end
    end
end

hybrid_mask=zeros([original_height original_width]);
hybrid_image=zeros([original_height original_width]);
for row=1:8:original_height
    for column=1:8:original_width
        if choosetype(1+fix(row/8),1+fix(column/8))==0;
            hybrid_mask(row:row+7,column:column+7)=zeros(8);
            hybrid_image(row:row+7,column:column+7)=...
                fractal_image(row:row+7,column:column+7);
        else
            hybrid_mask(row:row+7,column:column+7)=ones(8);
            hybrid_image(row:row+7,column:column+7)=...
                jpeg_image(row:row+7,column:column+7);
        end
    end
end

if DEBUG ~=0
    figure('Name','Hybrid Mask');
    imshow(hybrid_mask);
    imwrite(hybrid_mask,'mask.bmp');
    figure('Name','Hybrid Image');
    imshow(uint8(hybrid_image));
    hybrid_image_new=hybrid_image;
    imwrite(uint8(hybrid_image),'Hybridized Image.jpg','jpeg','Quality',...
    jpeg_tolerance);
    fid=fopen('Hybridized Image.jpg','r');
    hybridized_file=fread(fid);
    hybridized_bytes=length(hybridized_file);
    fclose(fid);
    hybrid_compression_ratio = original_file_size / hybridized_bytes;
end
hybrid_rms=findrms(double(original_image),double(hybrid_image));
hybrid_psnr=20*log10(255/hybrid_rms);

% Log the results...
fid=fopen('Compression Results.log','a');
```

```
fprintf(fid,'%2.2f %2.2f %2.3f %2.3f %2.3f %2.3f %2.3f %2.3f\n',mse_tolerance,...
    jpeg_tolerance,fractal_compression_ratio,fractal_psnr,...
    jpeg_compression_ratio,jpeg_psnr,hybrid_compression_ratio,hybrid_psnr);
fclose(fid);

timestamp=toc; %end stopwatch timer

%log the results to a central database
if MAKELOG ~=0
    fid=fopen(log_file,'a');
    if fid == -1
        fprintf(2,'unable to open log file\n');
        return
    end
    fprintf(fid,'run took %f seconds\n',timestamp);
    if (nargin >=1) && ischar(varargin{1})
        fprintf(fid,'image = %s\n',varargin{1});
    end
    fprintf(fid,'MSE tolerance = %f, JPEG quality = %f\n',...
        mse_tolerance, jpeg_tolerance);
    fprintf(fid,'fractal size = %i bytes, fractal PSNR = %f\n',...
        fractal_bytes, fractal_psnr);
    fprintf(fid,'JPEG size = %i bytes, JPEG PSNR = %f\n',...
        jpeg_bytes, jpeg_psnr);
    fprintf(fid,'hybrid size = %i bytes, hybrid PSNR = %f\n\n',...
        hybridized_bytes, hybrid_psnr);
    if (nargin >=1) && ischar(varargin{1})
        fprintf(fid,'"%s",%f,%f,%f,%f,%f,%f,%f\n',varargin{1},...
            mse_tolerance, fractal_bytes, fractal_psnr, jpeg_tolerance,...
            jpeg_bytes, jpeg_psnr, hybridized_bytes, hybrid_psnr);
    else
        fprintf(fid,'%f,%f,%f,%f,%f,%f,%f,%f\n',mse_tolerance,...
            fractal_bytes, fractal_psnr, jpeg_tolerance,...
            jpeg_bytes, jpeg_psnr, hybridized_bytes, hybrid_psnr);
    end
    fclose(fid);
end

fprintf(1,'Hybrid version %f\n',code_version);
fprintf(1,'run took %f seconds\n',timestamp);
if (nargin >=1) && ischar(varargin{1})
    fprintf(1,'image = %s\n',varargin{1});
end
fprintf(1,'MSE tolerance = %f, JPEG quality = %f\n',mse_tolerance,...
    jpeg_tolerance);
fprintf(1,'Fractal size = %i bytes, Fractal PSNR = %f\n',fractal_bytes,...
    fractal_psnr);
fprintf(1,'Fractal compression ratio = %f\n', fractal_compression_ratio);
fprintf(1,'JPEG size = %i bytes, JPEG PSNR = %f\n',jpeg_bytes, jpeg_psnr);
fprintf(1,'JPEG compression ratio = %f\n', jpeg_compression_ratio);
fprintf(1,'Hybrid size = %i bytes, Hybrid PSNR = %f\n',hybridized_bytes,...
    hybrid_psnr);
fprintf(1,'Hybrid compression ratio = %f\n\n', hybrid_compression_ratio);
if (nargin >=1) && ischar(varargin{1})
    fprintf(1,'"%s",%f,%f,%f,%f,%f,%f,%f\n\n',varargin{1},...
        mse_tolerance, fractal_bytes, fractal_psnr, jpeg_tolerance,...
        jpeg_bytes, jpeg_psnr, hybridized_bytes, hybrid_psnr);
else
    fprintf(1,'%f,%f,%f,%f,%f,%f,%f,%f\n\n',mse_tolerance,...
        fractal_bytes, fractal_psnr, jpeg_tolerance, jpeg_bytes,...
        jpeg_psnr, hybridized_bytes, hybrid_psnr);
end
return

%%%%%%%%%%%%%%%%%%%%%%%%%%%%%%%%%%%%%%%%%%%%%%%%%%%%%%%%%%%%%%%%%%%%%%%%%%
%subroutine to create a single transform                               %
%%%%%%%%%%%%%%%%%%%%%%%%%%%%%%%%%%%%%%%%%%%%%%%%%%%%%%%%%%%%%%%%%%%%%%%%%%
function [result_scale, result_mse, result_average]=...
    create_transform(image,row,column,range_size)
global SIZE_BITS
global AVERAGE_BITS
```

```
global SCALE_BITS
global one_transform
global mse_tolerance
result_transform=one_transform;
range=image(row:(row+range_size-1), column:(column+range_size-1));
%rbar=sum(sum(range))/(range_size*range_size);
rbar=mean2(range);
%grayscale test
measured_mse=findmse(range,rbar*ones(range_size));
if (measured_mse <= mse_tolerance)
    %scale factor = -1 means grayscale block
    result_scale=-1;
    result_average=rbar;
    result_mse=measured_mse;
else
    domain_size=fix(2*range_size);
    half_size=fix(range_size/2);
    domain_row=row-half_size;
    domain_column=column-half_size;
    domain_row=fix(min(max(domain_row,1),(size(image,1)-domain_size-1)));
    domain_column=fix(min(max(domain_column,1),(size(image,2)-...
        domain_size-1)));
     domain=(image(domain_row:2:(domain_row+domain_size-1),...
        domain_column:2:(domain_column+domain_size-1))+...
        image(domain_row+1:2:(domain_row+domain_size-1),...
        domain_column:2:(domain_column+domain_size-1)) +...
        image(domain_row:2:(domain_row+domain_size-1),...
        domain_column+1:2:(domain_column+domain_size-1))+...
        image(domain_row+1:2:(domain_row+domain_size-1),...
        domain_column+1:2:(domain_column+domain_size-1)) )/4;
    dbar=mean(mean(domain));
        rd=range.*domain;
    rrbar=mean2(range.*range);
    ddbar=mean2(domain.*domain);
    rdbar=mean2(range.*domain);
    rmean=fix(rbar/8)*8+4;
    rrsum=(rrbar-(2*rbar-rmean)*rmean);
    ddsum=(ddbar-dbar*dbar);
    rdsum=(rdbar-rbar*dbar);
    if ddsum == 0
        k=7;
    else
        k=floor((rdsum/ddsum)*8.0);
    end
    k=max(min(k,7),0);
    kk=0.0625+(k/8.0);
    measured_mse=sqrt(kk*kk*ddsum+rrsum-2.0*kk*rdsum);
    result_scale=k;
    result_mse=measured_mse;
        % result_average=8*floor(rbar/8)+4;
        result_average=max(0,min(255,round(rbar)));
end
return;

%%%%%%%%%%%%%%%%%%%%%%%%%%%%%%%%%%%%%%%%%%%%%%%%%%%%%%%%%%%%%%%%%%%%%%%%%%%%%%
%subroutine to measure MSE for a block                                     %
%%%%%%%%%%%%%%%%%%%%%%%%%%%%%%%%%%%%%%%%%%%%%%%%%%%%%%%%%%%%%%%%%%%%%%%%%%%%%%
function [measured_mse]=findmse(block1,block2)
global SIZE_BITS
global AVERAGE_BITS
global SCALE_BITS
global one_transform
global mse_tolerance
if size(block1) ~= size(block2)
    measured_mse=-1;
    return;
end;
measured_mse= (sum(sum((block1-block2).^2)))/(size(block1,1)*size(block1,2));
```

APPENDIX E

```
function [varargout]=Hybrid_2(varargin);
% A 3-way hybrid image compression algorithm based upon fractal, jpeg, and
% jpeg2000 methods.
%
% Original source code from searchless_r4 by William A. Stapleton
% Modified by Michael Alan McNees to include JPEG2000 image support.
%Run with no parameters for details.
close all;
clear all;
fclose all;

warning off all
code_version=2.10;
close all

DEBUG=1;
MAKELOG=1;
log_file='Hybrid_2.log';

if nargin < 1
    disp('Usage: [varargout]=searchless_r4(varargin)');
    disp(' ');
    disp('There can be up to 3 input arguments.');
    disp('1---the image to be compressed.');
    disp('2---the Mean Square Error(MSE) tolerance.');
    disp('3---the JPEG quality factor.');
    disp('4---the JPEG2000 explicit Quantization step size.');
    disp('If any parameters are omitted, the user will be prompted.');
end

%the following are general data definitions and structure assignments
global SIZE_BITS
global AVERAGE_BITS
global SCALE_BITS
global one_transform
global mse_tolerance

SIZE_BITS=3;      %number of bits to store size of range
AVERAGE_BITS=3; %minimum number of bits to use to store range average
SCALE_BITS=3;   %number of bits to store domain-to-range contrast scale
%MAX_SIZE_BITS=5;

one_transform = struct('size',8,'average',0.0,'mse',0.0,'filled',0,...
'scale',0.0);
temp_transform=one_transform;

%use open_image to select an image
%make sure selected image is correct form and compatible size

if nargin<1
    [raw_image original_file_size fname]=open_image;
    % Display the original file size in bytes...
    fprintf('Original FileSize: %d bytes.\n',original_file_size);
    figure('Name', 'Original Image');
    imshow(raw_image);
    jpeg2k_image=raw_image;
    if isempty(raw_image)
        disp('Cancel was selected...File Open Dialog Box...');
        close all
        return
    end
else
        if isgray(varargin{1})
        raw_image=varargin{1};
    elseif isrgb(varargin{1})
        if DEBUG ~=0
            disp('Warning, current algorithm is only defined for ');
            disp('grayscale images.');
            disp('Converting to grayscale');
        end
```

207

```
        raw_image=rgb2gray(varargin{1});
    elseif ischar(varargin{1})
        if DEBUG ~=0
            disp('Attempting to open ')
            disp(varargin{1})
            disp(' as image file');
        end
        raw_image=open_image(varargin{1});
    else
        disp('Do not know how to handle ')
        disp(varargin{1})
        disp(' as image');
        return
    end
end

if isrgb(raw_image)
    if DEBUG ~=0
        disp('Warning, current algorithm is only defined for grayscale');
        disp(' images.');
        disp('Converting to grayscale');
    end
    raw_image=rgb2gray(raw_image);
end

%prompt the user for a fractal mean-square error tolerance
if nargin < 2
    mse_input=inputdlg('Enter the MSE tolerance level','MSE Tolerance',...
        1,{'4.0'});
    if isempty(mse_input)    % Check to see if the user clicked Cancel...
        disp('Cancel was selected...Input Dialog Box... MSE tolerance...');
        return
    end
        mse_tolerance=str2num(mse_input{1});
else
    if isnumeric(varargin{2})
        mse_tolerance=varargin{2};
    elseif ischar(varargin{2})
        mse_tolerance=str2num(varargin{2});
    else
        disp('Do not know how to handle ');
        disp(varargin{2});
        disp(' as MSE tolerance');
        return
    end
end

% Create a JPEG version of the image.  Prompt the user for a JPEG
% Quality Tolerance value.

if nargin < 3
    jpeg_input=inputdlg('Enter the JPEG tolerance level',...
        'JPEG Tolerance',1,{'85'});
    if isempty(jpeg_input)
        disp('Cancel was selected...JPEG tolerance input Dialog Box...');
        return
    end
    jpeg_tolerance=str2num(jpeg_input{1});
else
    if isnumeric(varargin{3})
        jpeg_tolerance=varargin{3};
    elseif ischar(varargin{3})
        jpeg_tolerance=str2num(varargin{3});
    else
        disp('Do not know how to handle ');
        disp(varargin{3});
        disp(' as JPEG tolerance');
        return;
    end
end
```

```
% Create a JPEG2000 version of the image.  Prompt the user for an
% JPEG2000 Explicit Quantization level.

if nargin < 4
    jpeg2k_input=inputdlg('Enter the JPEG2k Explicit Quantization level',...
        'JPEG2k Quantization',1,{'4.5'});
    if isempty(jpeg2k_input)
        disp('Cancel was selected...JPEG2k Quantization input Dialog Box...');
        return
    end
    jpeg2k_quantization=str2num(jpeg2k_input{1});
else
    if isnumeric(varargin{4})
        jpeg2k_quantization=varargin{4};
    elseif ischar(varargin{4})
        jpeg2k_quantization=str2num(varargin{4});
    else
        disp('Do not know how to handle ');
        disp(varargin{4});
        disp(' as JPEG2k Quantization');
        return;
    end
end

tic; %start time)r

%adjust the image size so that the compressed image is an integer
%multiple of the largest range size
%the reconstructed image should
[raw_height raw_width]=size(raw_image);
MAX_SIZE_BITS = log2(raw_height /4);
MAX_RANGE_SIZE=2^MAX_SIZE_BITS;

original_height=MAX_RANGE_SIZE*floor((raw_height+(MAX_RANGE_SIZE-1))/...
    MAX_RANGE_SIZE);
original_width=MAX_RANGE_SIZE*floor((raw_width+(MAX_RANGE_SIZE-1))/...
    MAX_RANGE_SIZE);
raw_average=sum(sum(raw_image))/(raw_height*raw_width);
original_image=raw_average*ones(original_height,original_width);
original_image(1:raw_height,1:raw_width)=raw_image;
original_image=double(original_image);

%create a structure array to store the fractal data.
% transforms=repmat(one_transform,original_height,original_width);
size_bits=0;
AVERAGE_QUANTIZER=(2^AVERAGE_BITS+size_bits);
fractal_size=ones([original_height original_width]);
fractal_average=AVERAGE_QUANTIZER*floor((original_image+...
    (AVERAGE_QUANTIZER/2))/AVERAGE_QUANTIZER);
fractal_mse=zeros([original_height original_width]);
fractal_scale=-1*ones([original_height original_width]);
fractal_filled=ones([original_height original_width]);

if DEBUG ~=0
    disp('creating 2x2, 4x4, 8x8, ... structures');
end

for size_bits=1:MAX_SIZE_BITS
    range_size=fix(2^size_bits);
    if DEBUG ~=0
        disp(range_size);
    end
    half_size=fix(range_size/2);
%     current_mse_tolerance=mse_tolerance*sqrt(range_size/2);
    for row=1:range_size:original_height
        for column=1:range_size:original_width
            if fractal_mse(row,column) <= mse_tolerance && ...
                    fractal_mse(row,column+half_size) <= mse_tolerance && ...
                    fractal_mse(row+half_size,column) <= mse_tolerance && ...
                    fractal_mse(row+half_size,column+half_size) <= mse_tolerance
                %create transform for row,column at range_size
```
209

```
                [temp_scale temp_mse temp_average]=...
                    create_transform(original_image,row,column,range_size);
                if temp_mse <= mse_tolerance
                    fractal_scale(row,column)=temp_scale;
                    fractal_mse(row,column)=temp_mse;
                    fractal_average(row,column)=temp_average;
                    fractal_size(row,column)=range_size;
                    fractal_filled(row,column)=1;
                    fractal_filled(row,column+half_size)=0;
                    fractal_filled(row+half_size,column)=0;
                    fractal_filled(row+half_size,column+half_size)=0;
                end
            end
        end
    end
end

%generate an initial image for the fractal reconstruction from the stored
%average values for each range block.
fractal_image=128*ones([original_height,original_width]);
for row=1:original_height
    for column=1:original_width
        if fractal_filled(row,column) == 1
            fractal_image(row:(row+fractal_size(row,column)-1),...
                column:(column+fractal_size(row,column)-1))= ...
                fractal_average(row,column)*ones(fractal_size(row,column));
        end
    end
end

if DEBUG ~=0
    figure('Name','Fractal Image');
    imshow(uint8(fractal_image));
end

%reconstruction of image from fractal data
for iteration=1:5
    for row=1:original_height
        for column=1:original_width
            if fractal_filled(row,column) == 1
                if fractal_scale(row,column) == -1
                    new_image(row:row+fractal_size(row,column)-1,...
                        column:column+fractal_size(row,column)-1)=...
                        fractal_average(row,column)*...
                        ones(fractal_size(row,column));
                else
                    range_size=fractal_size(row,column);
                    domain_size=fix(2*range_size);
                    half_size=fix(range_size/2);
                    domain_row=row-half_size;
                    domain_column=column-half_size;
                    domain_row=fix(min(max(domain_row,1),...
                        (original_height-domain_size-1)));
                    domain_column=fix(min(max(domain_column,1),...
                        (original_width-domain_size-1)));
                    domain=(fractal_image(domain_row:2:...
                        (domain_row+domain_size-1),...
                        domain_column:2:(domain_column+domain_size-1))+...
                        fractal_image(domain_row+1:2:...
                        (domain_row+domain_size-1),domain_column:2:...
                        (domain_column+domain_size-1)) +...
                        fractal_image(domain_row:2:...
                        (domain_row+domain_size-1),...
                        domain_column+1:2:(domain_column+domain_size-1))+...
                        fractal_image(domain_row+1:2:...
                        (domain_row+domain_size-1),...
                        domain_column+1:2:(domain_column+domain_size-1)) )/4;
                    k= 0.0625+ (fractal_scale(row,column)/8.0);
                    new_image(row:row+range_size-1,...
                        column:column+range_size-1)=...
                        k*(domain-mean(mean(domain)))+...
```

```
                        fractal_average(row,column);
                end
            end
        end
    end
    fractal_image=new_image;
end

fractal_bits=0;
%calculate the size of the fractal file
for row=1:original_height
    for column=1:original_width
        if fractal_filled(row,column)==1
            if fractal_size(row,column)==1
                fractal_bits=fractal_bits+5;
            else
                fractal_bits=fractal_bits+9;
            end
        end
    end
end
fractal_bytes=ceil(fractal_bits/8)+15;
fractal_compression_ratio = original_file_size / fractal_bytes;
fractal_rms=findrms(double(original_image),double(fractal_image));
%fractal_rms=findrms(uint8(original_image),uint8(fractal_image));
fractal_psnr=20*log10(255/fractal_rms);


%*************************************************************************%
%                                                                       %
% Beginning of the JPEG image compression routine...                    %
%                                                                       %
%*************************************************************************%
imwrite(uint8(original_image),'searchless_temp.jpg','jpeg',...
    'Quality',jpeg_tolerance);
jpeg_image=open_image('searchless_temp.jpg');
fid=fopen('searchless_temp.jpg','r');
    if fid == -1
        fprintf(2,'unable to open file searchless_temp.jpg\n');
        return
    end
jpeg_file=fread(fid);
fclose(fid);
jpeg_bytes=length(jpeg_file);
jpeg_compression_ratio = original_file_size / jpeg_bytes;
jpeg_file=[];

if DEBUG ~=0
    figure('Name','JPEG Image');;
    imshow(jpeg_image);
end
jpeg_rms=findrms(double(original_image),double(jpeg_image));
%jpeg_rms=findrms(uint8(original_image),uint8(jpeg_image));
jpeg_psnr=20*log10(255/jpeg_rms);

%*************************************************************************%
%                                                                       %
% End of the JPEG image compression routine...                          %
%                                                                       %
%*************************************************************************%


%*************************************************************************%
%                                                                       %
% Beginning of the JPEG2000 image compression routine...                %
%                                                                       %
%*************************************************************************%
    qs=jpeg2k_quantization;
    oi=double(imread(fname));
    jpeg2k_exq = im2jpeg2k(uint8(oi), 1, [ qs qs qs qs ]);
    image2k_exq = jpeg2k2im(jpeg2k_exq);
```

```
    oi_bytes = bytes(oi);
    exq_jpeg2k_bytes = bytes(jpeg2k_exq);
    figure('Name','JPEG2000 Image---Explicit Quantization');
    imshow(jpeg2k2im(jpeg2k_exq));
    jpeg2k_exq_comp_ratio = oi_bytes / exq_jpeg2k_bytes;
    jpeg2k_rms_exq = compare(oi, image2k_exq);
    jpeg2k_psnr_exq = 20*log10(255/jpeg2k_rms_exq);

%***********************************************************************%
%                                                                       %
% End of the JPEG2000 image compression routine...                      %
%                                                                       %
%***********************************************************************%
choosetype=zeros([original_height original_width]/8);
for row=1:8:original_height
    for column=1:8:original_width
        fractal_mse=findmse(original_image(row:row+7,column:column+7),...
            fractal_image(row:row+7,column:column+7));
        jpeg_mse=findmse(original_image(row:row+7,column:column+7),...
            double(jpeg_image(row:row+7,column:column+7)));
        jpeg2k_mse=findmse(original_image(row:row+7,column:column+7),...
            double(image2k_exq(row:row+7,column:column+7)));
        if fractal_mse < jpeg_mse
            if fractal_mse < jpeg2k_mse
                choosetype(1+fix(row/8),1+fix(column/8))=0;
            else
                choosetype(1+fix(row/8),1+fix(column/8))=1;
            end
        end
        if jpeg2k_mse < fractal_mse
            if jpeg2k_mse < jpeg_mse
                choosetype(1+fix(row/8),1+fix(column/8))=2;
            end
        end
        if jpeg_mse < jpeg2k_mse
            if jpeg_mse < fractal_mse
                choosetype(1+fix(row/8),1+fix(column/8))=1;
            end
        end
    end
end
%choosetype
hybrid_mask=zeros([original_height original_width]);
hybrid_image=zeros([original_height original_width]);
for row=1:8:original_height
    for column=1:8:original_width
        if choosetype(1+fix(row/8),1+fix(column/8))==0;
            hybrid_mask(row:row+7,column:column+7)=zeros(8);
            hybrid_image(row:row+7,column:column+7)=...
                fractal_image(row:row+7,column:column+7);
        end
        if choosetype(1+fix(row/8),1+fix(column/8))==1;
            hybrid_mask(row:row+7,column:column+7)=ones(8);
            hybrid_image(row:row+7,column:column+7)=...
                jpeg_image(row:row+7,column:column+7);
        end
        if choosetype(1+fix(row/8),1+fix(column/8))==2;
            hybrid_mask(row:row+7,column:column+7)=ones(8);
            hybrid_image(row:row+7,column:column+7)=...
                image2k_exq(row:row+7,column:column+7);
        end
    end
end

if DEBUG ~=0
    figure('Name','Hybrid Mask');
    imshow(hybrid_mask);
    imwrite(hybrid_mask,'mask.bmp');
    figure('Name','Hybrid Image');
    imshow(uint8(hybrid_image));
    hybrid_image_new=hybrid_image;
```

```
        cl = im2jpeg2k(uint8(hybrid_image),1,[qs qs qs qs]);
        f1 = jpeg2k2im(cl);
        exq_jpeg2k_bytes = bytes(cl);
        imwrite(uint8(hybrid_image),'Hybridized Image.jpg','jpeg','Quality',jpeg_tolerance);
        fid=fopen('Hybridized Image.jpg','r');
        hybridized_file=fread(fid);
        hybridized_bytes=length(hybridized_file);
        fclose(fid);
        hybrid_compression_ratio = original_file_size / hybridized_bytes;
end
hybrid_rms=findrms(double(original_image),double(hybrid_image));
%hybrid_rms=findrms(uint8(original_image),uint8(hybrid_image));
hybrid_psnr=20*log10(255/hybrid_rms);

% Log the results...
fid=fopen('Compression Results.log','a');
fprintf(fid,'%2.2f %2.2f %2.3f %2.3f %2.3f %2.3f %2.3f %2.3f\n',mse_tolerance,...
    jpeg_tolerance,fractal_compression_ratio,fractal_psnr,...
    jpeg_compression_ratio,jpeg_psnr,hybrid_compression_ratio,hybrid_psnr);
fclose(fid);


timestamp=toc; %end stopwatch timer

%log the results to a central database
if MAKELOG ~=0
    fid=fopen(log_file,'a');
    if fid == -1
        fprintf(2,'unable to open log file\n');
        return
    end
    fprintf(fid,'run took %f seconds\n',timestamp);
    if (nargin >=1) && ischar(varargin{1})
        fprintf(fid,'image = %s\n',varargin{1});
    end
    fprintf(fid,'MSE tolerance = %f, JPEG quality = %f\n',...
        mse_tolerance, jpeg_tolerance);
    fprintf(fid,'fractal size = %i bytes, fractal PSNR = %f\n',...
        fractal_bytes, fractal_psnr);
    fprintf(fid,'JPEG size = %i bytes, JPEG PSNR = %f\n',...
        jpeg_bytes, jpeg_psnr);
    fprintf(fid,'hybrid size = %i bytes, hybrid PSNR = %f\n\n',...
        hybridized_bytes, hybrid_psnr);
    if (nargin >=1) && ischar(varargin{1})
        fprintf(fid,'"%s",%f,%f,%f,%f,%f,%f,%f,%f\n',varargin{1},...
            mse_tolerance, fractal_bytes, fractal_psnr, jpeg_tolerance,...
            jpeg_bytes, jpeg_psnr, hybridized_bytes, hybrid_psnr);
    else
        fprintf(fid,'%f,%f,%f,%f,%f,%f,%f,%f\n',mse_tolerance,...
            fractal_bytes, fractal_psnr, jpeg_tolerance,...
            jpeg_bytes, jpeg_psnr, hybridized_bytes, hybrid_psnr);
    end
    fclose(fid);
end

fprintf(1,'searchless_r4 version %f\n',code_version);
fprintf(1,'run took %f seconds\n',timestamp);
if (nargin >=1) && ischar(varargin{1})
    fprintf(1,'image = %s\n',varargin{1});
end
fprintf(1,'MSE tolerance = %f, JPEG quality = %f\n',mse_tolerance,...
    jpeg_tolerance);
fprintf(1,'Fractal size = %i bytes, Fractal PSNR = %f\n',fractal_bytes,...
    fractal_psnr);
fprintf(1,'Fractal compression ratio = %f\n', fractal_compression_ratio);
fprintf(1,'JPEG size = %i bytes, JPEG PSNR = %f\n',jpeg_bytes, jpeg_psnr);
fprintf(1,'JPEG compression ratio = %f\n', jpeg_compression_ratio);
fprintf(1,'Hybrid size = %i bytes, Hybrid PSNR = %f\n',hybridized_bytes,...
    hybrid_psnr);
fprintf(1,'Hybrid compression ratio = %f\n\n', hybrid_compression_ratio);
if (nargin >=1) && ischar(varargin{1})
```

```
        fprintf(1,'"%s",%f,%f,%f,%f,%f,%f,%f,%f\n\n',varargin{1},...
            mse_tolerance, fractal_bytes, fractal_psnr, jpeg_tolerance,...
            jpeg_bytes, jpeg_psnr, hybridized_bytes, hybrid_psnr);
else
    fprintf(1,'%f,%f,%f,%f,%f,%f,%f,%f\n\n',mse_tolerance,...
        fractal_bytes, fractal_psnr, jpeg_tolerance, jpeg_bytes,...
        jpeg_psnr, hybridized_bytes, hybrid_psnr);
end
return

%%%%%%%%%%%%%%%%%%%%%%%%%%%%%%%%%%%%%%%%%%%%%%%%%%%%%%%%%%%%%%%%%%%%%%%%%%
%subroutine to create a single transform                               %
%%%%%%%%%%%%%%%%%%%%%%%%%%%%%%%%%%%%%%%%%%%%%%%%%%%%%%%%%%%%%%%%%%%%%%%%%%
function [result_scale, result_mse, result_average]=...
    create_transform(image,row,column,range_size)
global SIZE_BITS
global AVERAGE_BITS
global SCALE_BITS
global one_transform
global mse_tolerance
result_transform=one_transform;
range=image(row:(row+range_size-1), column:(column+range_size-1));
%rbar=sum(sum(range))/(range_size*range_size);
rbar=mean2(range);
%grayscale test
measured_mse=findmse(range,rbar*ones(range_size));
if (measured_mse <= mse_tolerance)
    %scale factor = -1 means grayscale block
    result_scale=-1;
    result_average=rbar;
    result_mse=measured_mse;
else
    domain_size=fix(2*range_size);
    half_size=fix(range_size/2);
    domain_row=row-half_size;
    domain_column=column-half_size;
    domain_row=fix(min(max(domain_row,1),(size(image,1)-domain_size-1)));
    domain_column=fix(min(max(domain_column,1),(size(image,2)-...
        domain_size-1)));
      domain=(image(domain_row:2:(domain_row+domain_size-1),...
        domain_column:2:(domain_column+domain_size-1)+...
        image(domain_row+1:2:(domain_row+domain_size-1),...
        domain_column:2:(domain_column+domain_size-1) +...
        image(domain_row:2:(domain_row+domain_size-1),...
        domain_column+1:2:(domain_column+domain_size-1))+...
        image(domain_row+1:2:(domain_row+domain_size-1),...
        domain_column+1:2:(domain_column+domain_size-1)) )/4;
    dbar=mean(mean(domain));
      rd=range.*domain;
    rrbar=mean2(range.*range);
    ddbar=mean2(domain.*domain);
    rdbar=mean2(range.*domain);
    rmean=fix(rbar/8)*8+4;
    rrsum=(rrbar-(2*rbar-rmean)*rmean);
    ddsum=(ddbar-dbar*dbar);
    rdsum=(rdbar-rbar*dbar);
    if ddsum == 0
        k=7;
    else
        k=floor((rdsum/ddsum)*8.0);
    end
    k=max(min(k,7),0);
    kk=0.0625+(k/8.0);
    measured_mse=sqrt(kk*kk*ddsum+rrsum-2.0*kk*rdsum);
    result_scale=k;
    result_mse=measured_mse;
        % result_average=8*floor(rbar/8)+4;
        result_average=max(0,min(255,round(rbar)));
end
return;
```

214

```
%%%%%%%%%%%%%%%%%%%%%%%%%%%%%%%%%%%%%%%%%%%%%%%%%%%%%%%%%%%%%%%%%%%%%%%%%%
%subroutine to measure MSE for a block                                 %
%%%%%%%%%%%%%%%%%%%%%%%%%%%%%%%%%%%%%%%%%%%%%%%%%%%%%%%%%%%%%%%%%%%%%%%%%%
function [measured_mse]=findmse(block1,block2)
global SIZE_BITS
global AVERAGE_BITS
global SCALE_BITS
global one_transform
global mse_tolerance
if size(block1) ~= size(block2)
    measured_mse=-1;
    return;
end;
measured_mse= (sum(sum((block1-block2).^2)))/(size(block1,1)*size(block1,2));
```

APPENDIX F

```
function [picture original_file_size file_name]=open_image(varargin)
% Original source code provided by William A. Stapleton
% Modified by Michael Alan McNees to include the passing of the
% original file name as additional argument back to the calling program.
current_directory=cd;
if nargin<1,
    [file,path]=uigetfile({...
        '*.jpg;*.jpeg;*.bmp;*.gif;*.tif;*.tiff;*.png;*.pbm;*.pgm;*.ppm';...
        '*.pnm;*.pcx;*.ras;*.xwd;*.cur;*.ico','All Image Types';...
        '*.jpg;*.jpeg','JPEG (*.jpg, *.jpeg)';...
        '*.bmp','BMP (*.bmp)';...
        '*.gif','GIF (*.gif)';...
        '*.tif;*.tiff','TIFF (*.tif, *.tiff)';...
        '*.png','PNG (*.png)';...
        '*.pbm;*.pgm;*.ppm;*.pnm','PNM (*.pbm, *.pgm, *.ppm, *.pnm)';...
        '*.pcx','PCX (*.pcx)';...
        '*.ras','RAS (*.ras)';...
        '*.xwd','XWD (*.xwd)';...
        '*.cur;*.ico','Cursors and Icons (*.cur, *.ico)';...
        '*.*','All files (*.*)'});
    if isequal(file,0)|isequal(path,0)
        disp('File not found');
        picture=[];
        original_file_size=0;
        return
    end
    if strcmp(current_directory,path) == 0 & ~isequal(path,0)
        cd(path);
    end
else
    file=varargin{1};
end
picture=imread(file);
cd(current_directory);

size=imfinfo(file);
original_file_size=size.FileSize;
file_name=size.Filename;
```

APPENDIX G

```
function [c, s] = wavefast(x, n, varargin)
%WAVEFAST Perform multi-level 2-dimensional fast wavelet transform.
%   [C, L] = WAVEFAST(X, N, LP, HP) performs a 2D N-level FWT of
%   image (or matrix) X with respect to decomposition filters LP and
%   HP.
%
%   [C, L] = WAVEFAST(X, N, WNAME) performs the same operation but
%   fetches filters LP and HP for wavelet WNAME using WAVEFILTER.
%
%   Scale parameter N must be less than or equal to log2 of the
%   maximum image dimension.  Filters LP and HP must be even.  To
%   reduce border distortion, X is symmetrically extended. That is,
%   if X = [c1 c2 c3 ... cn] (in 1D), then its symmetric extension
%   would be [... c3 c2 c1 c1 c2 c3 ... cn cn cn-1 cn-2 ...].
%
%   OUTPUTS:
%      Matrix C is a coefficient decomposition vector:
%
%         C = [ a(n) h(n) v(n) d(n) h(n-1) ... v(1) d(1) ]
%
%      where a, h, v, and d are columnwise vectors containing
%      approximation, horizontal, vertical, and diagonal coefficient
%      matrices, respectively.  C has 3n + 1 sections where n is the
%      number of wavelet decompositions.
%
%      Matrix S is an (n+2) x 2 bookkeeping matrix:
%
%         S = [ sa(n, :); sd(n, :); sd(n-1, :); ... ; sd(1, :); sx ]
%
%      where sa and sd are approximation and detail size entries.
%
%   See also WAVEBACK and WAVEFILTER.

%   Copyright 2002-2004 R. C. Gonzalez, R. E. Woods, & S. L. Eddins
%   Digital Image Processing Using MATLAB, Prentice-Hall, 2004
%   $Revision: 1.5 $  $Date: 2003/10/13 01:14:17 $

% Check the input arguments for reasonableness.
error(nargchk(3, 4, nargin));

if nargin == 3
   if ischar(varargin{1})
      [lp, hp] = wavefilter(varargin{1}, 'd');
   else
      error('Missing wavelet name.');
   end
else
      lp = varargin{1};    hp = varargin{2};
end

fl = length(lp);     sx = size(x);

if (ndims(x) ~= 2) | (min(sx) < 2) | ~isreal(x) | ~isnumeric(x)
   error('X must be a real, numeric matrix.');
end

if (ndims(lp) ~= 2) | ~isreal(lp) | ~isnumeric(lp) ...
      | (ndims(hp) ~= 2) | ~isreal(hp) | ~isnumeric(hp) ...
      | (fl ~= length(hp)) | rem(fl, 2) ~= 0
   error(['LP and HP must be even and equal length real, ' ...
          'numeric filter vectors.']);
end

if ~isreal(n) | ~isnumeric(n) | (n < 1) | (n > log2(max(sx)))
   error(['N must be a real scalar between 1 and ' ...
          'log2(max(size((X))).']);
end

% Init the starting output data structures and initial approximation.
c = [];      s = sx;     app = double(x);
```

217

```
% For each decomposition ...
for i = 1:n
    % Extend the approximation symmetrically.
    [app, keep] = symextend(app, fl);

    % Convolve rows with HP and downsample. Then convolve columns
    % with HP and LP to get the diagonal and vertical coefficients.
    rows = symconv(app, hp, 'row', fl, keep);
    coefs = symconv(rows, hp, 'col', fl, keep);
    c = [coefs(:)' c];      s = [size(coefs); s];
    coefs = symconv(rows, lp, 'col', fl, keep);
    c = [coefs(:)' c];

    % Convolve rows with LP and downsample. Then convolve columns
    % with HP and LP to get the horizontal and next approximation
    % coeffcients.
    rows = symconv(app, lp, 'row', fl, keep);
    coefs = symconv(rows, hp, 'col', fl, keep);
    c = [coefs(:)' c];
    app = symconv(rows, lp, 'col', fl, keep);
end

% Append final approximation structures.
c = [app(:)' c];        s = [size(app); s];

%-------------------------------------------------------------------%
function [y, keep] = symextend(x, fl)
% Compute the number of coefficients to keep after convolution
% and downsampling. Then extend x in both dimensions.

keep = floor((fl + size(x) - 1) / 2);
y = padarray(x, [(fl - 1) (fl - 1)], 'symmetric', 'both');

%-------------------------------------------------------------------%
function y = symconv(x, h, type, fl, keep)
% Convolve the rows or columns of x with h, downsample,
% and extract the center section since symmetrically extended.

if strcmp(type, 'row')
    y = conv2(x, h);
    y = y(:, 1:2:end);
    y = y(:, fl / 2 + 1:fl / 2 + keep(2));
else
    y = conv2(x, h');
    y = y(1:2:end, :);
    y = y(fl / 2 + 1:fl / 2 + keep(1), :);
end
```

APPENDIX H

```
function x = huff2mat(y)
%HUFF2MAT Decodes a Huffman encoded matrix.
%   X = HUFF2MAT(Y) decodes a Huffman encoded structure Y with uint16
%   fields:
%     Y.min    Minimum value of X plus 32768
%     Y.size   Size of X
%     Y.hist   Histogram of X
%     Y.code   Huffman code
%
%   The output X is of class double.
%
%   See also MAT2HUFF.

%   Copyright 2002-2004 R. C. Gonzalez, R. E. Woods, & S. L. Eddins
%   Digital Image Processing Using MATLAB, Prentice-Hall, 2004
%   $Revision: 1.5 $  $Date: 2003/11/21 13:17:50 $

if ~isstruct(y) | ~isfield(y, 'min') | ~isfield(y, 'size') | ...
      ~isfield(y, 'hist') | ~isfield(y, 'code')
   error('The input must be a structure as returned by MAT2HUFF.');
end

sz = double(y.size);   m = sz(1);   n = sz(2);
xmin = double(y.min) - 32768;           % Get X minimum
map = huffman(double(y.hist));          % Get Huffman code (cell)

% Create a binary search table for the Huffman decoding process.
% 'code' contains source symbol strings corresponding to 'link'
% nodes, while 'link' contains the addresses (+) to node pairs for
% node symbol strings plus '0' and '1' or addresses (-) to decoded
% Huffman codewords in 'map'. Array 'left' is a list of nodes yet to
% be processed for 'link' entries.

code = cellstr(char('', '0', '1'));     % Set starting conditions as
link = [2; 0; 0];   left = [2 3];       % 3 nodes w/2 unprocessed
found = 0;   tofind = length(map);      % Tracking variables

while length(left) & (found < tofind)
   look = find(strcmp(map, code{left(1)}));    % Is string in map?
   if look                              % Yes
      link(left(1)) = -look;            % Point to Huffman map
      left = left(2:end);               % Delete current node
      found = found + 1;                % Increment codes found

   else                                 % No, add 2 nodes & pointers
      len = length(code);               % Put pointers in node
      link(left(1)) = len + 1;

      link = [link; 0; 0];              % Add unprocessed nodes
      code{end + 1} = strcat(code{left(1)}, '0');
      code{end + 1} = strcat(code{left(1)}, '1');

      left = left(2:end);               % Remove processed node
      left = [left len + 1 len + 2];    % Add 2 unprocessed nodes
   end
end

x = unravel(y.code', link, m * n);      % Decode using C 'unravel'
x = x + xmin - 1;                       % X minimum offset adjust
x = reshape(x, m, n);                   % Make vector an arra
```

219

www.ingramcontent.com/pod-product-compliance
Lightning Source LLC
LaVergne TN
LVHW042332060326
832902LV00006B/134